Teeth: A Very Short Introduction

VERY SHORT INTRODUCTIONS are for anyone wanting a stimulating and accessible way into a new subject. They are written by experts, and have been translated into more than 45 different languages.

The series began in 1995, and now covers a wide variety of topics in every discipline. The VSI library now contains over 500 volumes—a Very Short Introduction to everything from Psychology and Philosophy of Science to American History and Relativity—and continues to grow in every subject area.

Titles in the series include the following:

AFRICAN HISTORY John Parker and Richard Rathbone
AGEING Nancy A. Pachana
AGNOSTICISM Robin Le Poidevin
AGRICULTURE Paul Brassley and Richard Soffe
ALEXANDER THE GREAT Hugh Bowden
ALGEBRA Peter M. Higgins
AMERICAN HISTORY Paul S. Boyer
AMERICAN IMMIGRATION David A. Gerber
AMERICAN LEGAL HISTORY G. Edward White
AMERICAN POLITICAL HISTORY Donald Critchlow
AMERICAN POLITICAL PARTIES AND ELECTIONS L. Sandy Maisel
AMERICAN POLITICS Richard M. Valelly
THE AMERICAN PRESIDENCY Charles O. Jones
AMERICAN SLAVERY Heather Andrea Williams
THE AMERICAN WEST Stephen Aron
AMERICAN WOMEN'S HISTORY Susan Ware
ANAESTHESIA Aidan O'Donnell
ANARCHISM Colin Ward
ANCIENT EGYPT Ian Shaw
ANCIENT GREECE Paul Cartledge
THE ANCIENT NEAR EAST Amanda H. Podany
ANCIENT PHILOSOPHY Julia Annas
ANCIENT WARFARE Harry Sidebottom
ANGLICANISM Mark Chapman
THE ANGLO-SAXON AGE John Blair
ANIMAL BEHAVIOUR Tristram D. Wyatt
ANIMAL RIGHTS David DeGrazia
ANXIETY Daniel Freeman and Jason Freeman
ARCHAEOLOGY Paul Bahn
ARISTOTLE Jonathan Barnes
ART HISTORY Dana Arnold
ART THEORY Cynthia Freeland
ASTROPHYSICS James Binney
ATHEISM Julian Baggini
THE ATMOSPHERE Paul I. Palmer
AUGUSTINE Henry Chadwick
THE AZTECS Davíd Carrasco
BABYLONIA Trevor Bryce
BACTERIA Sebastian G. B. Amyes
BANKING John Goddard and John O. S. Wilson
BARTHES Jonathan Culler
BEAUTY Roger Scruton
THE BIBLE John Riches
BLACK HOLES Katherine Blundell
BLOOD Chris Cooper
THE BODY Chris Shilling
THE BOOK OF MORMON Terryl Givens
BORDERS Alexander C. Diener and Joshua Hagen
THE BRAIN Michael O'Shea
THE BRICS Andrew F. Cooper
BRITISH POLITICS Anthony Wright

BUDDHA Michael Carrithers
BUDDHISM Damien Keown
BUDDHIST ETHICS Damien Keown
BYZANTIUM Peter Sarris
CANCER Nicholas James
CAPITALISM James Fulcher
CATHOLICISM Gerald O'Collins
CAUSATION Stephen Mumford and Rani Lill Anjum
THE CELL Terence Allen and Graham Cowling
THE CELTS Barry Cunliffe
CHEMISTRY Peter Atkins
CHILD PSYCHOLOGY Usha Goswami
CHINESE LITERATURE Sabina Knight
CHOICE THEORY Michael Allingham
CHRISTIAN ART Beth Williamson
CHRISTIAN ETHICS D. Stephen Long
CHRISTIANITY Linda Woodhead
CIRCADIAN RHYTHMS Russell Foster and Leon Kreitzman
CITIZENSHIP Richard Bellamy
CLASSICAL MYTHOLOGY Helen Morales
CLASSICS Mary Beard and John Henderson
CLIMATE Mark Maslin
CLIMATE CHANGE Mark Maslin
COGNITIVE NEUROSCIENCE Richard Passingham
THE COLD WAR Robert McMahon
COLONIAL AMERICA Alan Taylor
COLONIAL LATIN AMERICAN LITERATURE Rolena Adorno
COMBINATORICS Robin Wilson
COMMUNISM Leslie Holmes
COMPLEXITY John H. Holland
THE COMPUTER Darrel Ince
COMPUTER SCIENCE Subrata Dasgupta
CONFUCIANISM Daniel K. Gardner
CONSCIOUSNESS Susan Blackmore
CONTEMPORARY ART Julian Stallabrass
CONTEMPORARY FICTION Robert Eaglestone
CONTINENTAL PHILOSOPHY Simon Critchley
CORAL REEFS Charles Sheppard
CORPORATE SOCIAL RESPONSIBILITY Jeremy Moon
COSMOLOGY Peter Coles
CRIMINAL JUSTICE Julian V. Roberts
CRITICAL THEORY Stephen Eric Bronner
THE CRUSADES Christopher Tyerman
CRYSTALLOGRAPHY A. M. Glazer
DADA AND SURREALISM David Hopkins
DANTE Peter Hainsworth and David Robey
DARWIN Jonathan Howard
THE DEAD SEA SCROLLS Timothy Lim
DECOLONIZATION Dane Kennedy
DEMOCRACY Bernard Crick
DEPRESSION Jan Scott and Mary Jane Tacchi
DERRIDA Simon Glendinning
DESIGN John Heskett
DEVELOPMENTAL BIOLOGY Lewis Wolpert
DIASPORA Kevin Kenny
DINOSAURS David Norman
DREAMING J. Allan Hobson
DRUGS Les Iversen
DRUIDS Barry Cunliffe
THE EARTH Martin Redfern
EARTH SYSTEM SCIENCE Tim Lenton
ECONOMICS Partha Dasgupta
EGYPTIAN MYTH Geraldine Pinch
EIGHTEENTH-CENTURY BRITAIN Paul Langford
THE ELEMENTS Philip Ball
EMOTION Dylan Evans
EMPIRE Stephen Howe
ENGLISH LITERATURE Jonathan Bate
THE ENLIGHTENMENT John Robertson
ENVIRONMENTAL ECONOMICS Stephen Smith
ENVIRONMENTAL POLITICS Andrew Dobson
EPICUREANISM Catherine Wilson
EPIDEMIOLOGY Rodolfo Saracci
ETHICS Simon Blackburn
THE ETRUSCANS Christopher Smith
EUGENICS Philippa Levine
THE EUROPEAN UNION John Pinder and Simon Usherwood
EVOLUTION Brian and Deborah Charlesworth

EXISTENTIALISM Thomas Flynn
THE EYE Michael Land
FAMILY LAW Jonathan Herring
FASCISM Kevin Passmore
FEMINISM Margaret Walters
FILM Michael Wood
FILM MUSIC Kathryn Kalinak
THE FIRST WORLD WAR Michael Howard
FOOD John Krebs
FORENSIC PSYCHOLOGY David Canter
FORENSIC SCIENCE Jim Fraser
FORESTS Jaboury Ghazoul
FOSSILS Keith Thomson
FOUCAULT Gary Gutting
FREE SPEECH Nigel Warburton
FREE WILL Thomas Pink
FREUD Anthony Storr
FUNDAMENTALISM Malise Ruthven
FUNGI Nicholas P. Money
THE FUTURE Jennifer M. Gidley
GALAXIES John Gribbin
GALILEO Stillman Drake
GAME THEORY Ken Binmore
GANDHI Bhikhu Parekh
GEOGRAPHY John Matthews and David Herbert
GEOPOLITICS Klaus Dodds
GERMAN LITERATURE Nicholas Boyle
GERMAN PHILOSOPHY Andrew Bowie
GLOBAL CATASTROPHES Bill McGuire
GLOBAL ECONOMIC HISTORY Robert C. Allen
GLOBALIZATION Manfred Steger
GOD John Bowker
GRAVITY Timothy Clifton
THE GREAT DEPRESSION AND THE NEW DEAL Eric Rauchway
HABERMAS James Gordon Finlayson
THE HEBREW BIBLE AS LITERATURE Tod Linafelt
HEGEL Peter Singer
HERODOTUS Jennifer T. Roberts
HIEROGLYPHS Penelope Wilson
HINDUISM Kim Knott
HISTORY John H. Arnold
THE HISTORY OF ASTRONOMY Michael Hoskin
THE HISTORY OF CHEMISTRY William H. Brock
THE HISTORY OF LIFE Michael Benton
THE HISTORY OF MATHEMATICS Jacqueline Stedall
THE HISTORY OF MEDICINE William Bynum
THE HISTORY OF TIME Leofranc Holford-Strevens
HIV AND AIDS Alan Whiteside
HOLLYWOOD Peter Decherney
HUMAN ANATOMY Leslie Klenerman
HUMAN EVOLUTION Bernard Wood
HUMAN RIGHTS Andrew Clapham
THE ICE AGE Jamie Woodward
IDEOLOGY Michael Freeden
INDIAN PHILOSOPHY Sue Hamilton
THE INDUSTRIAL REVOLUTION Robert C. Allen
INFECTIOUS DISEASE Marta L. Wayne and Benjamin M. Bolker
INFINITY Ian Stewart
INFORMATION Luciano Floridi
INNOVATION Mark Dodgson and David Gann
INTELLIGENCE Ian J. Deary
INTERNATIONAL MIGRATION Khalid Koser
INTERNATIONAL RELATIONS Paul Wilkinson
IRAN Ali M. Ansari
ISLAM Malise Ruthven
ISLAMIC HISTORY Adam Silverstein
ISOTOPES Rob Ellam
ITALIAN LITERATURE Peter Hainsworth and David Robey
JESUS Richard Bauckham
JOURNALISM Ian Hargreaves
JUDAISM Norman Solomon
JUNG Anthony Stevens
KABBALAH Joseph Dan
KANT Roger Scruton
KNOWLEDGE Jennifer Nagel
THE KORAN Michael Cook
LATE ANTIQUITY Gillian Clark
LAW Raymond Wacks
THE LAWS OF THERMODYNAMICS Peter Atkins
LEADERSHIP Keith Grint
LEARNING Mark Haselgrove

LEIBNIZ Maria Rosa Antognazza
LIBERALISM Michael Freeden
LIGHT Ian Walmsley
LINGUISTICS Peter Matthews
LITERARY THEORY Jonathan Culler
LOCKE John Dunn
LOGIC Graham Priest
MACHIAVELLI Quentin Skinner
MAGIC Owen Davies
MAGNA CARTA Nicholas Vincent
MAGNETISM Stephen Blundell
MARINE BIOLOGY Philip V. Mladenov
MARTIN LUTHER Scott H. Hendrix
MARTYRDOM Jolyon Mitchell
MARX Peter Singer
MATERIALS Christopher Hall
MATHEMATICS Timothy Gowers
THE MEANING OF LIFE Terry Eagleton
MEASUREMENT David Hand
MEDICAL ETHICS Tony Hope
MEDIEVAL BRITAIN John Gillingham and Ralph A. Griffiths
MEDIEVAL LITERATURE Elaine Treharne
MEDIEVAL PHILOSOPHY John Marenbon
MEMORY Jonathan K. Foster
METAPHYSICS Stephen Mumford
MICROBIOLOGY Nicholas P. Money
MICROECONOMICS Avinash Dixit
MICROSCOPY Terence Allen
THE MIDDLE AGES Miri Rubin
MILITARY JUSTICE Eugene R. Fidell
MINERALS David Vaughan
MODERN ART David Cottington
MODERN CHINA Rana Mitter
MODERN FRANCE Vanessa R. Schwartz
MODERN IRELAND Senia Pašeta
MODERN ITALY Anna Cento Bull
MODERN JAPAN Christopher Goto-Jones
MODERNISM Christopher Butler
MOLECULAR BIOLOGY Aysha Divan and Janice A. Royds
MOLECULES Philip Ball
MOONS David A. Rothery
MOUNTAINS Martin F. Price
MUHAMMAD Jonathan A. C. Brown
MUSIC Nicholas Cook
MYTH Robert A. Segal
THE NAPOLEONIC WARS Mike Rapport
NELSON MANDELA Elleke Boehmer
NEOLIBERALISM Manfred Steger and Ravi Roy
NEWTON Robert Iliffe
NIETZSCHE Michael Tanner
NINETEENTH-CENTURY BRITAIN Christopher Harvie and H. C. G. Matthew
NORTH AMERICAN INDIANS Theda Perdue and Michael D. Green
NORTHERN IRELAND Marc Mulholland
NOTHING Frank Close
NUCLEAR PHYSICS Frank Close
NUMBERS Peter M. Higgins
NUTRITION David A. Bender
THE OLD TESTAMENT Michael D. Coogan
ORGANIC CHEMISTRY Graham Patrick
THE PALESTINIAN-ISRAELI CONFLICT Martin Bunton
PANDEMICS Christian W. McMillen
PARTICLE PHYSICS Frank Close
THE PERIODIC TABLE Eric R. Scerri
PHILOSOPHY Edward Craig
PHILOSOPHY IN THE ISLAMIC WORLD Peter Adamson
PHILOSOPHY OF LAW Raymond Wacks
PHILOSOPHY OF SCIENCE Samir Okasha
PHOTOGRAPHY Steve Edwards
PHYSICAL CHEMISTRY Peter Atkins
PILGRIMAGE Ian Reader
PLAGUE Paul Slack
PLANETS David A. Rothery
PLANTS Timothy Walker
PLATE TECTONICS Peter Molnar
PLATO Julia Annas
POLITICAL PHILOSOPHY David Miller
POLITICS Kenneth Minogue
POPULISM Cas Mudde and Cristóbal Rovira Kaltwasser
POSTCOLONIALISM Robert Young

POSTMODERNISM Christopher Butler
POSTSTRUCTURALISM Catherine Belsey
PREHISTORY Chris Gosden
PRESOCRATIC PHILOSOPHY Catherine Osborne
PRIVACY Raymond Wacks
PSYCHIATRY Tom Burns
PSYCHOLOGY Gillian Butler and Freda McManus
PSYCHOTHERAPY Tom Burns and Eva Burns-Lundgren
PUBLIC ADMINISTRATION Stella Z. Theodoulou and Ravi K. Roy
PUBLIC HEALTH Virginia Berridge
QUANTUM THEORY John Polkinghorne
RACISM Ali Rattansi
REALITY Jan Westerhoff
THE REFORMATION Peter Marshall
RELATIVITY Russell Stannard
RELIGION IN AMERICA Timothy Beal
THE RENAISSANCE Jerry Brotton
RENAISSANCE ART Geraldine A. Johnson
REVOLUTIONS Jack A. Goldstone
RHETORIC Richard Toye
RISK Baruch Fischhoff and John Kadvany
RITUAL Barry Stephenson
RIVERS Nick Middleton
ROBOTICS Alan Winfield
ROMAN BRITAIN Peter Salway
THE ROMAN EMPIRE Christopher Kelly
THE ROMAN REPUBLIC David M. Gwynn
RUSSIAN HISTORY Geoffrey Hosking
RUSSIAN LITERATURE Catriona Kelly
THE RUSSIAN REVOLUTION S. A. Smith
SAVANNAS Peter A. Furley
SCHIZOPHRENIA Chris Frith and Eve Johnstone
SCIENCE AND RELIGION Thomas Dixon
THE SCIENTIFIC REVOLUTION Lawrence M. Principe
SCOTLAND Rab Houston
SEXUALITY Véronique Mottier
SHAKESPEARE'S COMEDIES Bart van Es
SIKHISM Eleanor Nesbitt
SLEEP Steven W. Lockley and Russell G. Foster
SOCIAL AND CULTURAL ANTHROPOLOGY John Monaghan and Peter Just
SOCIAL PSYCHOLOGY Richard J. Crisp
SOCIAL WORK Sally Holland and Jonathan Scourfield
SOCIALISM Michael Newman
SOCIOLOGY Steve Bruce
SOCRATES C. C. W. Taylor
SOUND Mike Goldsmith
THE SOVIET UNION Stephen Lovell
THE SPANISH CIVIL WAR Helen Graham
SPANISH LITERATURE Jo Labanyi
SPORT Mike Cronin
STARS Andrew King
STATISTICS David J. Hand
STUART BRITAIN John Morrill
SYMMETRY Ian Stewart
TAXATION Stephen Smith
TELESCOPES Geoff Cottrell
TERRORISM Charles Townshend
THEOLOGY David F. Ford
TIBETAN BUDDHISM Matthew T. Kapstein
THE TROJAN WAR Eric H. Cline
THE TUDORS John Guy
TWENTIETH-CENTURY BRITAIN Kenneth O. Morgan
THE UNITED NATIONS Jussi M. Hanhimäki
THE U.S. CONGRESS Donald A. Ritchie
THE U.S. SUPREME COURT Linda Greenhouse
THE VIKINGS Julian Richards
VIRUSES Dorothy H. Crawford
VOLTAIRE Nicholas Cronk
WAR AND TECHNOLOGY Alex Roland
WATER John Finney
WILLIAM SHAKESPEARE Stanley Wells
WITCHCRAFT Malcolm Gaskill
THE WORLD TRADE ORGANIZATION Amrita Narlikar
WORLD WAR II Gerhard L. Weinberg

Peter S. Ungar

TEETH

A Very Short Introduction

Great Clarendon Street, Oxford, OX2 6DP,
United Kingdom

Oxford University Press is a department of the University of Oxford. It furthers the University's objective of excellence in research, scholarship, and education by publishing worldwide. Oxford is a registered trade mark of Oxford University Press in the UK and in certain other countries

First Edition published in 2014

Published in the United States of America by Oxford University Press
198 Madison Avenue, New York, NY 10016, United States of America

British Library Cataloguing in Publication Data
Data available

Library of Congress Control Number: 2013953522

ISBN 978–0–19–967059–8

Printed and bound by
CPI Group (UK) Ltd, Croydon, CR0 4YY

Contents

Acknowledgements

I thank the many mentors, colleagues, and students from whom I have learned what I know about teeth. I thank especially those that read over and commented on bits and pieces of this book and the editorial staff at Oxford University Press for their helpful suggestions and encouragement. Finally, and most importantly, I am grateful to my wife and daughters for their continued tolerance, patience, and understanding.

List of illustrations

Chapter 1
Teeth matter

'Whoa, look at those teeth, they're so cool!' I enjoy walking through the exhibits at natural history museums when I visit for research. This time it was the Smithsonian. The little girl, six or seven, dragged her younger brother by the arm across a crowded hall to see the skull of *Dimetrodon*, a mammal-like reptile that lived nearly 300 million years ago. Its teeth *are* cool—but so are yours. Think about it. Your teeth are the product of half a billion years of evolution. They provide fuel for the body by breaking apart other living things; and they do it again and again over a lifetime without themselves being broken in the process. It's like a perpetual death match in the mouth, with plants and animals developing tough or hard tissues for protection, and teeth evolving ways to sharpen or strengthen themselves to overcome those defences.

Why are we drawn to teeth in the halls of natural history museums and in picture books of fossil species? There's something visceral about them. Perhaps it's because our early ancestors spent so much time running away from teeth. Or maybe it's because they define us. As George Cuvier, the great 19th-century naturalist is often quoted to have said, 'Show me your teeth, I will tell you who you are.' We know intuitively something about an animal by looking at its mouth. Think of *Tyrannosaurus rex*, with its long, sharp teeth for killing prey and ripping flesh. A little closer to home, a recent survey of nearly 5,500 American singles by the online dating service Match.com found teeth to be the #1 attribute both men and women use to judge potential partners. Yes, teeth matter.

I have spent my entire adult life studying teeth, but what I *really* care about is how Nature works, how life came to be as it is today, and where we humans fit into the picture. Teeth matter to me because they are great tools for working these things out.

The ecological angle

Teeth can help us understand ecology, the study of how living things interact with one another and with their physical environment. What could be more fundamental to those interactions than feeding? An organism eats its neighbour for the fuel and raw materials needed to grow, sustain itself, and reproduce. Teeth matter because they are right in the middle of it, mediating between eater and eaten. They are the front line in Nature's 'struggle for existence', as Darwin called it.

Conventional wisdom suggests that teeth gave early vertebrates an edge in the 'arms race' between predator and prey. The filter feeding, jawless fishes that dominated the seas for hundreds of millions of years had no chance once jaws and teeth evolved. As the renowned 20th-century paleontologist Edwin Colbert wrote, 'A vertebrate without jaws was efficient after a fashion, but unless it became adapted to certain very specialized habits it was not well enough equipped for survival in a world where a pair of upper and lower jaws had evolved as a food-gathering mechanism.' In reality, though, jawless fishes did just fine for nearly 100 million years, much of it alongside their jawed and ultimately toothed cousins.

But teeth must have given those that had them an advantage for capturing and immobilizing prey. They could be used to scrape, pry, grasp, and nip all manner of living thing. And better access to nutrients meant more offspring, and more evolutionary success. Teeth spread quickly through the oceans of the early Palaeozoic Earth, whether toothed fishes sidelined their toothless cousins or not. As 20th-century paleontologist James Marvin Weller wrote, 'Although teeth rarely excite the attention that their importance warrants, their evolution among the early vertebrates without a doubt played an unrivaled role in the successful adaptation of these animals and their achievement of rapid and effective dominance in the organic world.'

The next major milestone for teeth was the ability to occlude, wherein opposing surfaces come together in a precise way for chewing. This evolved in some amphibians and reptiles, but today mammals own occlusion and chewing. They use these mechanisms to rupture plant cell walls and insect exoskeletons for access to nutrients that would otherwise pass through the gut undigested. Chewing also leads to smaller particles for swallowing, and more exposed surface area for digestive enzymes to act on. In other words, it means the extraction of more fuel and raw materials from a mouthful of food.

This is especially important for mammals because they are endotherms—they heat their bodies from within. It takes fuel, and lots of it, to be endothermic and keep the home furnace burning. Chewing gives mammals the energy needed to be active not only during the day but also the cool night, and to live in colder climates or places with more fluctuating temperatures. It allows them to sustain higher levels of activity and travel speeds to cover larger distances, avoid predators, capture prey, and make and care for offspring. Mammals are able to live in an incredible variety of habitats, from Arctic tundra to Antarctic pack ice, deep open waters to high-altitude mountaintops, and rainforests to deserts, in no small measure because of their teeth.

The paleontological angle

Teeth matter a lot to paleontologists. First, they are the most common vertebrate fossils we find, and many species are known only from their teeth. Second, we can use them to infer the diets of past animals because tooth size, shape, structure, wear, and chemistry all relate to what an animal eats. Because diet is such an important key to ecology, connecting tooth to food can help us reconstruct paleoecology, the relationships between past organisms and their environments. We can trace changes in these relationships over time if we have a reasonable fossil record. And if we combine this with models of past climates, we might even figure out how environmental change triggered evolution. We can begin to understand how animals in the past differed from or were similar to those alive today, and how present-day animals, including us, came to be the way we are.

A very brief introduction to the history of dental research

People have been thinking about teeth for a very long time. Aristotle discussed them in *De partibus animalium*, around 350 BC. His comparisons of tooth number, size, and shape among animals according to their diets served as the pinnacle of knowledge on teeth for nearly two millennia. And much of his work still stands the test of time. There are other bits and pieces on teeth that survive from antiquity, recorded in general works on anatomy and medicine by Hippocrates and Galen and their later compilations by Avicenna. But there is little else until Aristotle's corpus and other classical works began to spread through Europe as movable-type book printing took off in the late 15th and early 16th centuries. This prompted many studies on anatomy and zoology, several of which touched on teeth.

The first known book on teeth is *Artzney Buchlein* (1530), a short, anonymous collection of descriptions of dental pathologies and their treatments. Soon after, the great Flemish anatomist Andreas Vesalius devoted a chapter of *De humani corporis fabrica* (1542) to teeth. Another renowned anatomist of the time, Bartolomeo Eustacio, followed with *Libellus de dentibus* (1563), the first known book devoted in its entirety to dental structure and function. And Eustacio took a comparative approach, contrasting human teeth with those of other animals.

The invention of the microscope in the 17th century led to significant advances in understanding how teeth are put together. Antony van Leeuwenhoek and Marcello Malpighi documented the microscopic structure, or histology, of dental tissues in detail. And there were many more influential works in the 18th century, such as *Le chirugien dentiste* (1728) by Pierre Fauchard, and *The Natural History of the Human Teeth* (1771) by John Hunter.

But the golden era of odontography, the descriptive study of teeth, really came in the early 19th century. We owe much of our knowledge today to naturalists of the time, including Georges Cuvier, Richard Owen, and Christoph Giebel. And as the theory of natural selection began to take hold later in the century, comparative anatomists such as Thomas Henry Huxley, William Flower, and Richard Lydekker entered the mix. Work on dental histology also flourished. Some students might recognize Andres Retzius, Victor von Ebner, Samuel Salter, and John and Charles Tomes from the microscopic dental structures that bear their names. The transition from description to explanation, or odontography to odontology, also began late in the 19th century, with Edward Drinker Cope, Henry Fairfield Osborn, and others who developed models for the origin and evolution of teeth. These models are in large part still with us today.

Studies of teeth have continued to progress in the 20th century and into the 21st. Their results are the subject of this book. Works on growth and development have brought new insights, and we are beginning to discover the genetic controls over tooth size and shape. Recent advances in our understanding of how species relate to one another provide a framework for understanding how teeth evolve. Studies of tooth size, shape, structure, wear, and chemistry offer fresh insights into how teeth work, how animals use them today, and how they used them in the past. And new fossil finds are filling important gaps in the paleontological record, allowing us to document the major milestones in the evolution of teeth and chewing.

Chapter 2
Types and parts of teeth

Many different kinds of teeth are found in the animal kingdom. Not only do they vary between species, but also within a mouth, both by generation (baby teeth versus adult ones) and by type (front teeth versus back ones).

Variation within a mouth

Tooth generations. Most vertebrates shed and replace their teeth. Sharks, for example, can do it hundreds of times, with tens of thousands of teeth passing through the mouth in a lifetime. And replacements can differ in size, shape, and structure from their predecessors. Larger teeth often succeed smaller ones as the jaw grows throughout life in non-mammalian vertebrates. These replacements tend to come in an alternating pattern, every other or every third position to prevent large gaps in the tooth row.

Mammals do things differently because our jaws stop growing in adulthood. We don't need multiple generations of progressively larger teeth—two generations work just fine. Our baby teeth, also called milk or deciduous, are usually smaller, with thinner and whiter enamel; and their crowns and roots are shaped differently from those of their adult, or permanent, replacements. We replace our first twenty teeth, all but our molars, and add a dozen of those as space in the jaws is made available. The final molars erupt around the same time that jaw growth finishes. Most other mammalian species have a different pattern, though. Many are born with permanent teeth in place; their deciduous ones never completely form, or they erupt and are shed in the womb. And a few apparently keep their milk teeth and never replace them. Mice, for example, have adult teeth at birth, and toothed whales evidently never get them.

Tooth types. Many vertebrates are *homodont*; their teeth all look about the same. They are often cone- or needle-shaped, and function to acquire, capture, contain, or kill. Mammals and other animals that need to chew and break food into pieces are usually *heterodont*; their front and back teeth differ, with a dental division of labour for food acquisition and processing respectively. The sheepshead fish, for example, has front teeth that look like our incisors, used for scraping and grasping. Their back teeth are flat, pebble-like structures, used to crush sea urchins and other hard foods. Herbivorous lizards, like the iguana, also have dental differentiation, with cone-shaped front teeth for cropping vegetation, and more complex ones behind them for shredding. And mammals take this differentiation to an extreme, with four distinct types: incisors, canines, premolars, and molars (see Figure 1).

Incisors are the front teeth. These are usually flattened and shovel-shaped with one cusp and one root, but they may have more. Incisors serve a bevy of functions, from grasping or nipping to stripping, scraping, and other behaviours that bring food into the mouth in chunks small enough to chew or swallow. These teeth can be quite specialized, ranging from ever-growing chisels in rodents and rabbits used for gnawing, to comb-like structures in colugos with prongs for grooming, and tusks in elephants and narwhals used as tools for prying and digging, as weapons, or as special sensory organs.

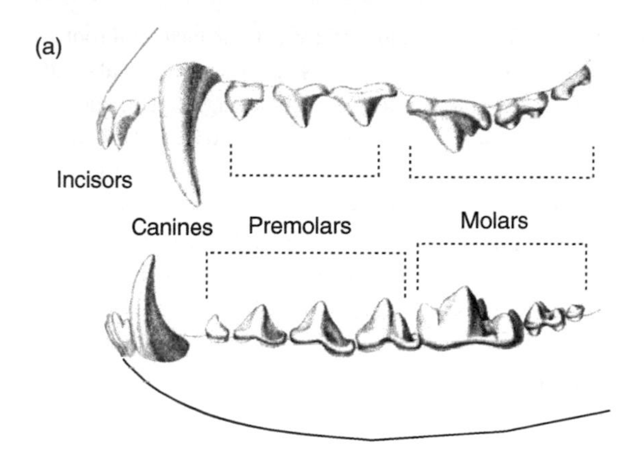

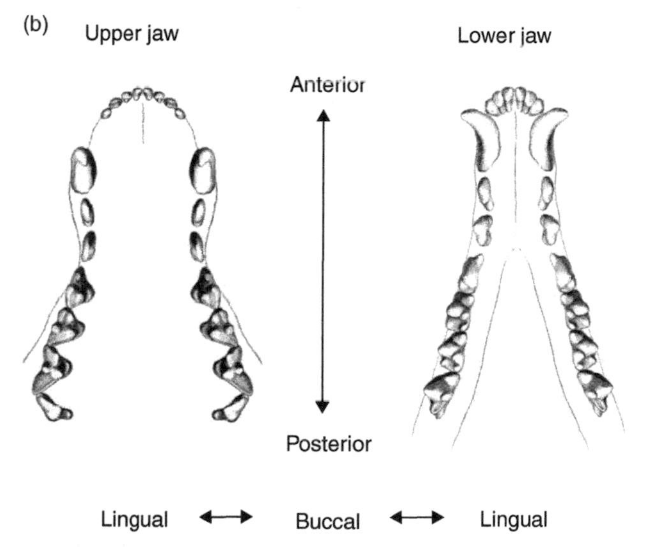

1. Tooth types and positions. A, fox teeth in side view; and B, quoll (an Australian marsupial) teeth of the upper (left) and lower (right) jaws

Canines are next. They are also usually a single cusp and root. They are long and dagger-like in some species, such as cats and many monkeys, with sharp, pointed edges for fighting, or stabbing, biting, and holding prey. They are small and incisor-like in others, such as moles and many herbivores, and act along with the other front teeth in ingestion. Some are modified into tusks, as in the walrus, hippopotamus, and razorback boar. The canines and incisors together are called the anterior teeth, distinguished from those behind, which are the postcanines or cheek teeth.

The premolars are just behind the canines. These vary from slight, single-cusped teeth in shrews, to more elaborate ones for crushing in hyenas, slicing in cats, grinding in antelopes, etc. These also often vary along the row, grading from canine-like to molar-like. Ours typically have two cusps, which is why dentists call them bicuspids. Several possums, rat kangaroos, and their kin have premolars shaped a bit like a steak knife, each with a single long and thin serrated blade.

Molars are the back teeth. These vary from pegs in dolphins, aardvarks, and sloths, to complex, intricate structures with multiple bumps, crests, and grooves, in capybaras, horses, and elephants. These are used, along with premolars, to fracture and fragment food into smaller parts by shearing, crushing, and grinding.

Researchers often distinguish mammals on the basis of the number of each tooth type, expressed as a dental formula. The dental formula for our permanent teeth is I2/2, C1/1, P2/2, M3/3—two upper and two lower incisors, one upper and one lower canine, two upper and two lower premolars, and three upper and three lower molars on each side of the mouth. Upper and lower numbers are included because they differ in many mammals. There is no need to separate left from right jaws because, except for the São Tomé collared fruit bat and the narwhal, they are mirror images of one another.

The ancestral marsupial and placental mammalian dental formulas are I5/4, C1/1, P3/3, M4/4, and I3/3, C1/1, P4/4, M3/3 respectively. Most today have fewer teeth, but some have more—spinner dolphins have up to 260 in the mouth at once.

Variation between mouths

To compare teeth between species we need some terms to describe their parts. For the anterior teeth, the front side is *labial*, the back side is *lingual*, toward the midline is *mesial*, and away from it is *distal*. For the postcanine teeth, the front is *anterior*, back is *posterior*, tongue side is *lingual*, and cheek side is *buccal* (*bucca* is Latin for cheek). We can describe a molar that is long front to back and narrow side to side as buccolingually compressed and anteroposteriorly elongate. To be fair, not all toothed animals have cheeks, and researchers that study non-mammals often refer to the sides of marginal teeth (the equivalent of cheek teeth) as external and internal. In this case, we can refer to teeth that are narrow side to side as mediolaterally compressed—medial and lateral being toward and away from the midline of the body respectively.

The biting end of a tooth is called the *occlusal* surface. This is where things can get complicated. The occlusal surface can have dozens of cusps, crests, and other features with long and intimidating names such as *hypoconulid* and *postmetacrista*. As the dental research guru Percy Butler lamented, 'Every student of comparative tooth morphology has first to overcome the rather considerable obstacle of a complicated nomenclature. This gives the impression that the subject is much more abstruse than it really is.' But the terminology isn't really that bad if you understand the rationale behind it. To get there, we need to step back to the 19th century and visit Edward Drinker Cope and his younger colleague, Henry Fairfield Osborn.

The Cope–Osborn model. Cope spent much of the 1870s and 1880s developing a model of progression from simple, primitive teeth to the complex, specialized ones of modern mammals such as cats and horses. Osborn then filled in many of the details, and developed the terms we use today to describe the parts of the crown. They believed that the mammalian upper tooth began as a simple cone-shaped structure, which Osborn called the *protocone.* They argued that a *paracone* and *metacone* formed in front of and behind the protocone respectively. According to the Cope–Osborn model, the paracone and metacone became displaced buccally over evolutionary time, and the protocone was offset lingually, resulting in a triangular structure called the *trigon.* A fourth cusp, the *hypocone,* was then added behind the protocone on a low shelf, or heel, called the *talon* (see Figure 2).

Cope and Osborn thought the lower molar evolved the same way, and Osborn used the suffix '-id' to distinguish lower from upper cusps. So, the *protoconid* was the original lower tooth cusp, with the *paraconid* added in front and the *metaconid* behind. In this case, though, the paraconid and metaconid were offset lingually and the protoconid was pushed to the buccal side of the crown. The trigon and opposing *trigonid* thus formed 'reversed triangles'. As with the upper tooth, a low shelf called the *talonid* evolved behind the trigonid, this time with up to three cusps: the *entoconid* on the lingual side, the *hypoconid* on the buccal side, and the *hypoconulid* on the back end.

Osborn named other features on the crown by combining the prefixes of nearby cusps with suffixes that indicate the feature type. Crests have the suffix *crista* or *cristid,* so the preprotocrista connects the paracone to the protocone. Secondary cusps often end with *conule* and *conulid.* The paraconule is next to the paracone. If the secondary cusp is on the *cingulum* or *cingulid,* which is a collar of enamel running along the sides of the crown, it has the suffix *style* or *stylid.* In fact, when the cingulum

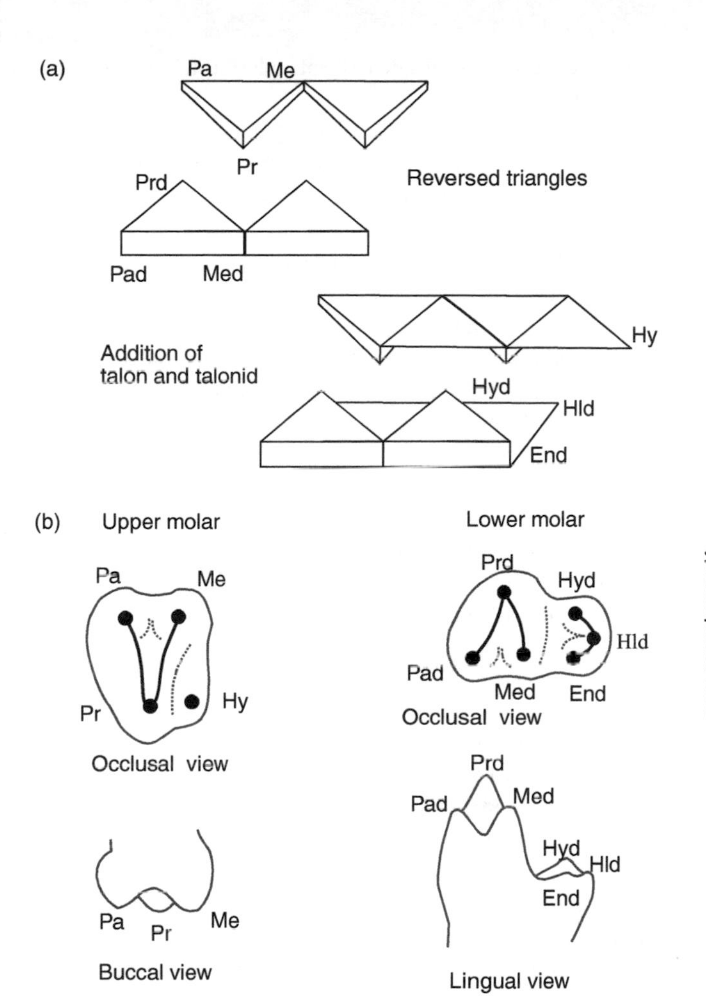

2. The Cope–Osborn model and cusp names. A, reverse triangle configuration of upper and lower molar teeth (above) and addition of talon and talonid (below); and B, upper and lower molar in occlusal and side views. Cusp names for the upper teeth are: Pa = paracone, Me = metacone, Pr = protocone, Hy = hypocone. Cusp names for lower teeth are: Pad = paraconid, Med = metaconid, Prd = protoconid, End = entoconid, Hyd = hypoconid, and Hld = hypoconulid

is expanded on the buccal side to form a platform, it is called the *stylar shelf*.

The naming system, then, is not so bad after all. But there is a problem. As the teeth of more fossil mammals were found during the 20th century, it became clear that the Cope–Osborn model was wrong. For example, the original cusp that Osborn called the protocone is actually the paracone on the trigon of later mammalian teeth. The original paracone became one of the stylar cusps (we call it *stylar cusp B*). To make matters worse, upper and lower molar cusps with the same prefixes today do not even match up. Stylar cusp B corresponds to the paraconid, and the paracone matches the protoconid.

Needless to say, Cope and Osborn's errors led to pandemonium as researchers struggled to update names to keep up with our improved understanding of how mammalian teeth evolve. But as Percy Butler noted, aptly and succinctly, 'Language is for communication.' The old names are too entrenched in the literature to abandon, and the least confusing solution is to stick with them, acknowledging that they do not mean what Osborn thought they did. I prefer to name things based on consistent location; so the inside back cusps on upper molars of a kangaroo and a monkey are hypocones, whether they came from the same structures in a common ancestor or not.

The tribosphenic molar. Cope and Osborn may have been wrong about how the mammalian ancestor got to the reversed triangle form, but that form itself and models for how today's molars evolved from it have held up better. Think of opposing rows of identical triangles arranged side by side with their tips pointing lingually for the uppers and buccally for the lowers. Imagine the rows are offset so opposing teeth fit between one another when they interdigitate, or come into occlusion. The sides of the lowers slide along those of the uppers like scissor blades when the jaw is raised and teeth brought together. We can attach low shelves, or

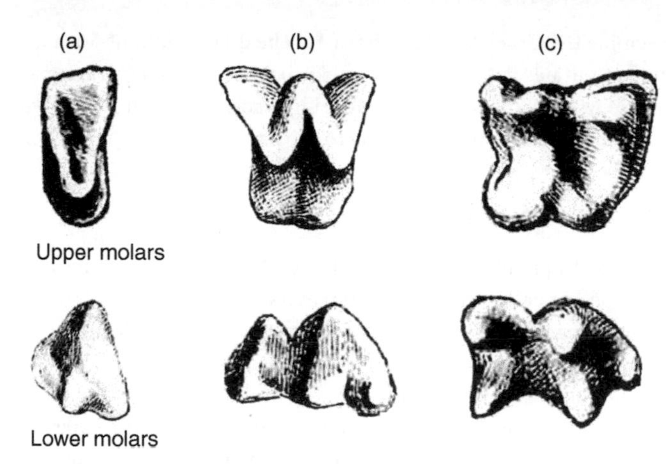

3. Variants of the tribosphenic molar. Upper and lower teeth. A, zalambdodont type; B, dilambdodont type; C, euthemorphic type

heels, to the backs of both the upper and lower triangles to face the opposing triangle and add a crushing component. The triangles represent the trigon and trigonid, and the shelves represent the talon and talonid. The eminent paleontologist George Gaylord Simpson coined the term *tribosphenic* (combining the Greek *triben*, to rub, with *sphen*, or wedge) in the 1930s to recognize the dual shearing and crushing functions of these teeth.

There are several real-life variants on this tribosphenic theme; the most common are called *euthemorphic*, *zalambdodont*, and *dilambdodont* (see Figure 3). Euthemorphic molars are fairly flat, or *bunodont*, and squared off, with a well-developed paracone and protocone in front, and a metacone and hypocone behind them. The zalambdodont form has its trigon pushed to the lingual edge of the tooth, with a well-developed paracone, but small or absent protocone and metacone. The buccal side is dominated by a broad stylar shelf with crests connecting the paracone with the parastyle and metastyle. These crests together are called the *ectoloph*, and resemble the Greek letter lambda (Λ). The dilambdodont form is similar, but has a second pair of crests behind the first (forming a double Λ, or W if you prefer), with ridges connecting the metastyle to the metacone to the mesostyle to the paracone to the parastyle.

What lies beneath

While teeth are often complex at the surface, there's even more going on beneath. Think about the extraordinary feat of engineering. Your teeth must concentrate and transmit the forces needed to break foods over and over again, up to millions of times over a lifetime, without being broken in the process. And they must do this with the raw materials that Nature has to offer—the very same ones used to make the plants and animals being eaten. What gives teeth their remarkable strength? The answer is a complex, composite structure evolved over half a billion years.

While the basic types of tissue that make up teeth vary among vertebrates, most mammals have enamel, dentine, cementum, and pulp. The principal tissue and basic skeleton of the tooth is dentine, with the crown covered by a cap of enamel, and roots by thin layers of cementum. The dentine core is hollow with the crown's interior chamber and the root canals connected, housing pulp, nerves, and blood vessels. Pulp is largely soft connective tissue, whereas enamel, dentine, and cementum are harder, made up of varying proportions of mineral (mostly hydroxyapatite, a form of calcium phosphate), along with organic matter and water. The relative proportions of these parts, their structures, and their distributions determine the strength of a tooth (see Figure 4).

Material scientists use very precise terms to describe properties such as strength, and some of their vocabulary is useful for considering interactions between tooth and food. We can start with *stress*, which is force per unit area. Stress can be increased by

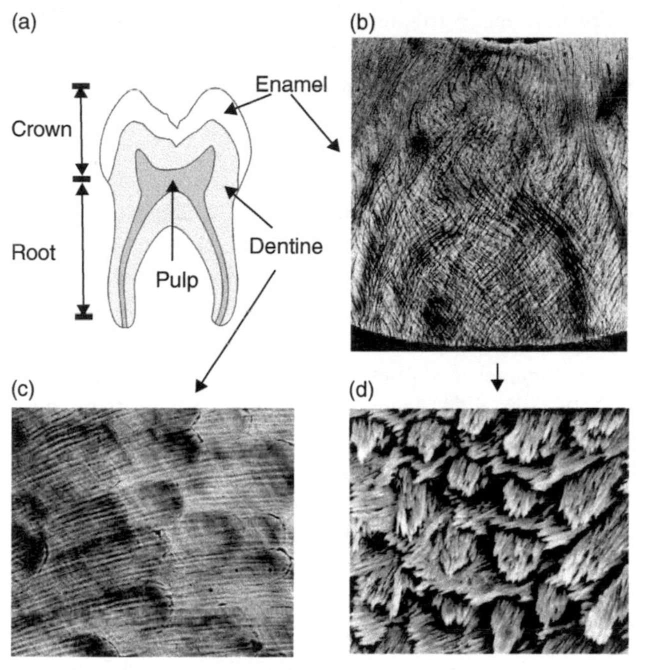

4. Tooth structure. A, section through an idealized tooth; B, section through enamel showing wiggling of prisms; and C, section through dentine showing tubules; D, magnified view of enamel prisms

increasing force, or by decreasing the area over which it is applied. That's why it takes less force to pound a sharp nail than a blunt one. A related term is *strain*, how much an object is deformed when a stress is applied to it: in simplest terms, it's change in length relative to the original length of that object. *Stiff* materials require more stress to generate the strain needed to cause *failure*. Failure can involve permanent deformation, fracture, or both. The stress at which an object begins to deform permanently or crack is called its *yield strength* or *fracture strength* respectively. Resistance to failure is often called *hardness*. Once a crack starts, resistance to its spread is called *fracture toughness*. Tough tissues require more energy to spread a crack than do *brittle* ones. We can also speak of hardness and toughness of teeth and food as their resistance to crack formation and propagation respectively.

Enamel is the hardest tissue in your body. It is about 97 per cent mineral by weight, with an elegant and ingenious microscopic structure designed by Nature for strength. Think of a wooden pencil. It breaks easily if bent along its length, but not if pressed inward from tip to eraser. Now think of thousands of pencils bundled together like bunches of dried spaghetti strands into large, cylindrical rods, and thousands of these rods packed together. Enamel crystallites are like pencils, 0.04 microns (0.000004 cm) thick. They are bunched together into prisms averaging about 5 microns across. Prisms are, in turn, packed together into layers. These prisms, and the crystallites within them, run the thickness of the enamel cap, from where it meets the dentine (the enamel–dentine junction, or EDJ) to the surface.

But when prisms run parallel and straight from the EDJ to the surface (called *radial* enamel), cracks can spread along the boundaries between them. These are stopped by wriggling, weaving, and twisting prisms around one another (called *decussation*). The change of direction drains energy available to a spreading crack. Layers of packed prisms can also be oriented in different ways and woven together to further toughen enamel, and some teeth even have layers of layers.

How does enamel achieve this complex microstructure? Cells called *ameloblasts* are packed into sheets in a developing tooth bud, or germ. These sheets migrate outward from what will be the EDJ, leaving a matrix of protein, mineral, and water behind them. Envision squeezing a bunch of open tubes of toothpaste together. If you were to move the tubes away from the direction of the flowing toothpaste as you squeeze, you'd be left with a cluster of parallel toothpaste trails resembling packed enamel prisms.

Jiggling the tubes as you move them would produce decussation. Ameloblasts are like these toothpaste tubes, starting at the EDJ, and moving outward to the eventual surface of the tooth, leaving enamel matrix behind them. After this, the cells absorb the water and organic bits, and pump in more mineral. Once the completed tooth erupts in the mouth, the ameloblasts are shed from its surface. This is why you cannot repair or replace your tooth enamel; the cells that secrete it are lost.

The rate of enamel matrix secretion and mineralization fluctuates throughout the day. This leads to alternating bulges and constrictions (imagine squeezing bundled toothpaste tubes in pulses, increasing and decreasing the force rhythmically), which results in daily incremental lines that cross-cut the developing sheet of prisms. We can use these to study fossils. These lines, called *cross striations*, are in a way like tree rings. They can be counted to determine how long it takes for a tooth cap to develop. And like tree rings, cross striations vary with stress. Dental researchers can even identify the day of birth, stressful for both mother and baby, by a well-defined *neonatal line*. Enamel formation rate also varies according to a roughly weekly cycle, resulting in distinctive incremental lines called *striae of Retzius*. These form ridges, or *perikymata*, where they intersect the surface.

Prism formation, layout, and packing patterns vary greatly among species. And, except for the agamid lizard, only mammals have enamel formed from prisms. Further, most primitive fishes do not even have enamel proper; instead they have a different highly mineralized tissue called *enameloid*, which forms from both ameloblasts and the cells that secrete dentine, *odontoblasts*. Enameloid-based teeth are still very strong. In fact, shark enameloid is as hard as our enamel, largely because the fluoroapatite mineral in enameloid is harder than the hydroxyapatite in our enamel. Also, fluoroapatite crystals interdigitate with protein fibres in the underlying dentine to strengthen the tooth further.

Dentine is a yellowish, bone-like tissue that is not as hard as enamel (only about 70 per cent mineral). It combines hydroxyapatite with protein (collagen) fibres, making it tough and elastic. Dentine is dominated by tiny tubes, or tubules (tens of thousands per mm^2), that run parallel from the EDJ inward. These house projections of odontoblasts, which, again, are the cells that secrete dentine. The odontoblast cell bodies are attached to the wall of the pulp chamber. After the tooth forms, dentine secretion slows, but continues as *secondary* dentine, which can nearly fill a pulp chamber over the course of a lifetime. And *tertiary* dentine forms to repair the tissue in reaction to irritation of the pulp. The different kinds of dentine vary not only in their timing of secretion and distribution, but also in their microscopic structure and chemical composition.

Dentine forms in a manner similar in some ways to enamel. A single layer of odontoblasts begin at what will become the EDJ. Instead of moving outward like ameloblasts, though, odontoblasts move inward, toward the eventual pulp chamber, leaving trails of collagen-rich predentine. The predentine forms in thin tubules that enclose the developing cell processes. Mineralization follows as in enamel formation, though in this case the organic component is not removed. Dentine has daily and longer-period incremental lines, similar to those of enamel, called *von Ebner* and *Andresen lines* respectively.

Cementum is slightly less mineralized than dentine, about 65 per cent mineral on average, with the rest organic matter and water. Cementum varies between species, in both thickness and distribution. It usually coats the root, and occasionally the crown, in mammals and some reptiles. As with dentine, a collagen-rich *precementum* forms first, with mineralization following. Cementum develops in two distinct layers. The deeper *intermediate cementum* is only about 100th of a millimetre thick, but it is rich in calcium and very hard. Intermediate cementum covers the root and seals its dentine tubules. The outer *dental cementum* is thicker but softer. It has bundles of collagen fibre, some of which remain unmineralized to attach to the *periodontal ligament*, which anchors the tooth in the jaw. Sheets of cementoblasts continue to lay down cementum throughout life, allowing for continuous reattachment of this ligament.

Like enamel and dentine, cementum has incremental lines, though the intervals at which they occur is the subject of some debate. In some cases at least, they reflect seasonal variation in metabolism, with two distinct lines per year.

Pulp. While pulp lacks the direct and obvious connection that the mineralized tissues have to tooth stress and strain, no review of dental structure is complete without considering it. Pulp is a soft, gelatinous connective tissue inside the crown chamber and root canal. It is made up of several layers, some containing cells that maintain the tissue, and others encasing blood vessels and nerve fibres that enter the roots through their tip or from tiny canals along their sides.

How teeth are made

The study of individual tissues offers hints on how teeth are made, but to consider the tooth as a whole and how Nature makes teeth different from one another, we need to delve deeper into developmental biology.

The embryos of complex animals, from flatworms to humans, divide into three layers early in development: the endoderm, mesoderm, and ectoderm. Vertebrates have a fourth layer, the neural crest, formed from the ectoderm. Our teeth begin to develop about six weeks after conception, starting with a band of ectodermal tissue called the *dental lamina*. During the first stage of development, the bud stage, cells grow from the lamina to form tooth buds, which push into the developing jaw, specifically, a layer of tissue derived from the neural crest called *ectomesenchyme* (see Figure 5).

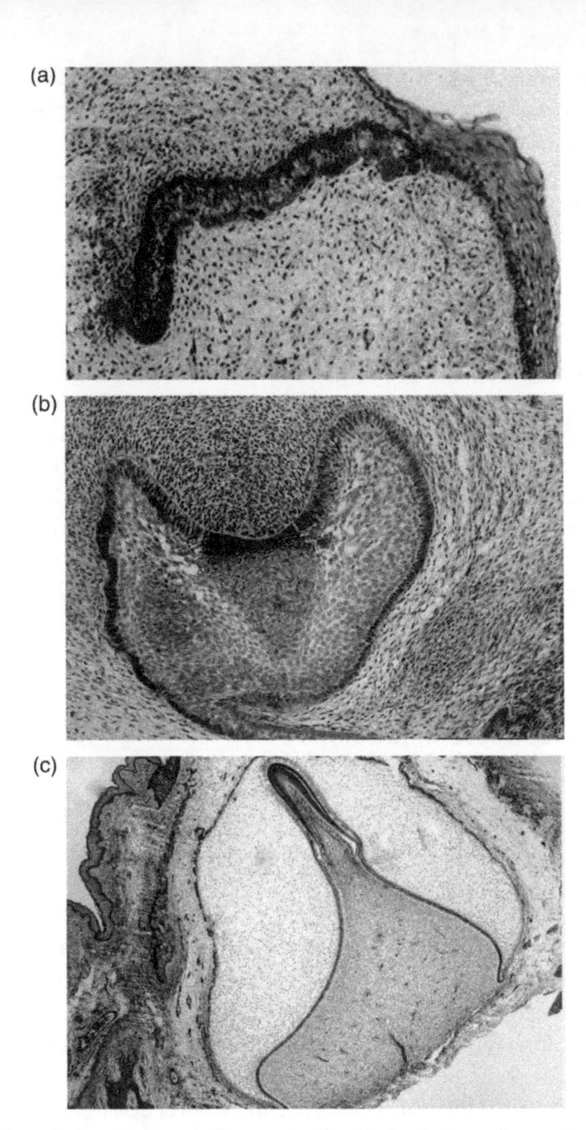

5. Dental development. Human teeth at A, bud stage; B, cap stage; C, bell stage

Next, during the cap stage, the bud grows to form a cap-shaped structure, the *enamel organ*. And ectomesenchyme just below the cap condenses into a mass called the *dental papilla*. Cells made from the enamel organ will ultimately form the enamel, and those from the papilla will make dentine and pulp. The membrane separating the two is the future site of the EDJ, again, where enamel meets dentine. Ectomesenchyme also forms a sack, or capsule, called the *dental follicle*, around the dental organ. Follicle cells make cementoblasts, alveolar bone around the teeth, and the periodontal ligament, which anchors tooth to jaw. The enamel organ, dental papilla, and dental follicle together comprise the developing tooth germ.

Finally, during the bell stage, the enamel organ starts to look a bit like a church bell as the underlying papilla presses deep into it. The bell differentiates into various tissues, and the tooth begins to take shape as they fold. The enamel organ produces, among other things, the ameloblasts. The dental papilla also differentiates, with layers ultimately forming odontoblasts and pulp tissues. Enamel and dentine then start forming the crown, from cusp tip toward the root. A layer of cells form a sheath that guides odontoblasts to cover the developing root canal with dentine. Cementoblasts then cover the root with cementum, and the alveolar bone and periodontal ligament form.

But how does Nature make teeth with different shapes? Developmental biologist Jukka Jernvall and his colleagues have been working for years on answering this question, bringing us ever closer to understanding how genes control differences in tooth form between species. Tooth crowns have different shapes because of how their tissues fold and grow where ectoderm meets ectomesenchyme during development. Different patterns mean different numbers, shapes, and placements of cusps and other features on a crown.

The process starts with a small concentration of cells at the tip of the tooth bud called the *primary enamel knot*. This is like a command centre that sends out signals using protein molecules. These signalling molecules start and stop cell division. But things get more complicated when a tooth has more than one cusp. The process requires secondary knots at the sites of the future cusp tips. These secondary knots are also signalling centres, making molecules that start and stop cell division and, ultimately, control crown shape. The sizes of individual cusps depend on timing; larger ones tend to begin developing before smaller ones. What about number and placement of cusps? Just before the primary knot dies, early in the cap stage, it sends out activator proteins to tell the tooth germ to make the secondary knots, and inhibitor ones to prevent them from forming. It becomes a race as the signals spread across the developing tissue. The distances between secondary knots depend on how much more quickly the activator proteins move. The ultimate number of cusps on a tooth is limited by these distances and space available on the crown.

This new understanding of how dental features form has very important implications for those of us that study teeth. First, instructions for making crown shape follow what is called a *cascading pattern*. The initial signals touch off a chain reaction, like a line of falling dominoes. Set them up, push the first one, and stand back. More importantly, because knots beget knots, there is no one-to-one relationship between genes and cusps. There is no 'hypocone gene', and cusps can no longer be considered independent and separate players in the story of dental evolution. In other words, we can no longer think about them the way Cope and Osborn did in the 19th century.

Chapter 3
What teeth do: food and feeding

Aristotle wrote around 350 BC, 'Teeth have one invariable office, namely the reduction of food.' We can't appreciate what teeth do without first understanding what vertebrates need from them—the energy and raw materials to live, grow, and reproduce.

The biospheric buffet

Woody Allen describes Nature in *Love and Death* as 'big fish eating little fish, and plants eating plants, and animals eating an . . . '. He tails off, but then continues: 'It's like an enormous restaurant, the way I see it.' We can think of the biosphere, the part of our planet that harbours life, as a kind of giant buffet, with animals picking and choosing different items with which to fill their plates. Because energy can come from carbohydrates, lipids, or proteins, and many nutrients can be synthesized by the body or absorbed from microbes living in the gut, animals have many options.

Animals can eat nutrient producers, such as green plants. Plants make the building blocks—simple sugars, amino acids, and fatty acids—from water, air, and rock. Then plants assemble them into complex carbohydrates, proteins, and lipids as needed. Herbivores break the complex molecules in plants back down to their basic parts during digestion, and then put them back together in different ways to meet their own needs. Carnivores eat the herbivores, and again ingest, digest, and reassemble the parts. And finally, decomposers get nutrients from the waste or corpses of plants or other animals. In this way, the basic chemicals of life are recycled, over and over again.

With so many options available, how do animals choose their diets? It's a matter of balancing costs and benefits. Some foods, such as grass on the savanna, may be abundant and easy to gather, but difficult to digest. Others, such as most animals, may be easy to digest, but rare, and hard to capture. There's also competition, and the need to eat while avoiding being eaten. Different animals make different choices to meet their needs for energy and raw materials. These choices lead to the challenges that teeth face.

Nutrient requirements. Carbohydrates are used generally for fuel. Most come from green plants, which take energy from sunlight to make sugar and oxygen from carbon dioxide and water. Plants bond simple sugars together, up to thousands at a time, to form chains called polysaccharides for structural support and energy storage. Cellulose is the most common. There is a lot of energy stored in cellulose, but vertebrates have difficulty accessing it. Only simple sugars can be absorbed directly by our guts, and we do not make the enzymes needed to break the bonds that hold them together. There are other sources of energy (lipids and proteins), but for most vertebrates, simple sugars, especially glucose, are important for powering the brain and a few other tissues. Animals can get glucose directly from the foods they eat, make it from other organic molecules, or host microorganisms in their guts that can cleave it from complex carbohydrates.

Lipids are another important source of energy, better suited to storage for later use. Most are simple chemical compounds that combine fatty acids with alcohol. When molecules of the alcohol glycerol, for example, are combined with three fatty acids, they form triglycerides, the main constituents of plant oils and animal fats. These are broken down in the digestive tract, but can reform once they cross the intestinal barrier. Lipids provide fuel and insulation, but also give structural support to cell membranes and regulate many cellular functions. While vertebrates can synthesize most of the fatty acids they need, there are some essential ones (for us, linoleic acid and linolenic acid) that must be acquired from food.

Proteins are also important sources of energy. These are long chains of amino acids that are disassembled so their parts can be burned for energy, used to make sugar, or converted to fat for storage. Proteins play many important roles in the body, from structural support to transport, and from directing chemical reactions to defence. Their function depends on the number and sequence of amino acids that make up the chain. There are 20 different types of amino acid in our proteins. These are strung together like necklaces made from different-coloured beads, and often folded into complex three-dimensional structures. Vertebrates can make about half the amino acids they need, but the others, the essential amino acids, must come from foods eaten, absorbed from gut bacteria, or cannibalized from existing proteins in the body.

Vitamins are other organic compounds needed for the body to function normally, grow, and reproduce. They are usually divided into two types: fat soluble, which can be stored by the body, and water soluble, which cannot. Most vertebrates need to consume, or obtain from gut bacteria, thirteen to seventeen different vitamins, defined in part by the inability of the body to synthesize them, at least in sufficient quantities. While we do not make vitamin C, many other organisms do. For them, ascorbate is not considered a vitamin. When animals plan their meals, they need to bear in mind that consumption of vitamins involves a delicate balancing act, as too much, especially of fat-soluble vitamins, can be toxic.

The body also requires *inorganic elements*, and at least twenty-two and perhaps up to forty or more play a role in normal metabolic function. These are divided into macrominerals and microminerals (or trace elements), which are distinguished by quantity needed. A convenient threshold is 50 mg/kg body weight or 100 mg/kg of food. Inorganic elements serve many functions in the body, from structural support to transport, regulation, etc. As with vitamins, though, minerals must be consumed in appropriate quantities to strike a balance between deficiency and toxicity.

The final nutrient is *water*. This is often overlooked, which is surprising given that it typically makes up more than half of our mass and 99 per cent of our molecules. Water is important for transport, metabolism, and temperature control. It is a major component of all bodily fluids and acts as a solvent and diluent. Water is constantly lost by the body, and needs to be replenished regularly. It can come from drinking and moisture in food, or it can be synthesized from other nutrients.

Diet categories. So where do vertebrates get these nutrients? Plants produce most of the energy in the biosphere, and there are about 300,000 land plant species alone from which herbivores can choose. But there are costs and challenges to eating many of these. With a couple of notable exceptions, such as fleshy fruits, plants tend to not want their parts eaten, and so have developed a bevy of defences to protect themselves. Plants produce about 33,000 chemical compounds that can harm a herbivore, and they make lignin and other substances to stiffen, harden, toughen, or otherwise discourage consumption. Still, herbivores have a powerful incentive to thwart these defences. Plants contain many nutrients, and their potential energy yield is staggering; cell walls are dominated by cellulose molecules, each of which can contain thousands of glucose units.

That's where gut symbiotes, bacteria in the lining of the gastrointestinal tract, come in. In fact, you have about ten times the number of gut microorganisms as cells in your body. These help prevent infection by pathogens, trigger normal developmental processes such as growth of epithelial cells, blood vessels, and lymphoid tissue, and break down complex carbohydrates that enter the gut. They also synthesize other nutrients, including fatty acids, amino acids, and vitamins. This helps herbivores meet their needs without their having to eat all the nutrients they cannot themselves make. And some have developed specialized anatomy to concentrate and keep microbes in the foregut, the hindgut, or both. Hindgut fermenters often have complex sacs or pouches in their large intestines, and pass heaps of low-quality food through their guts quickly. Some eat faeces or cecotropes (the so-called 'night droppings' well known to rabbit owners), and pass food through a second time to complete digestion. Hindgut fermenters range from mice to elephants, koalas to howling monkeys, and horses to rhinoceroses. Foregut fermenters, in contrast, often have complex, multi-chambered stomachs that can slow or restrict food passage to give items more time to ferment. Kangaroos and wallabies, colobus monkeys, hippopotamuses and peccaries, camels and llamas, sloths, and baleen whales are all foregut fermenters. In none of these animals, though, has foregut fermentation reached the level of aesthetic beauty achieved by the ruminants. Cows, deer, and their kin have four stomach chambers and regurgitate food to be chewed a second time for more complete digestion.

Another way to divide herbivores is by whether they are grazers that eat grass, browsers that eat parts of higher-growing plants such as shrubs and trees, or mixed feeders that eat a combination of the two types. Grasses tend to have thick cell walls dominated by cellulose, but they are rich in complex carbohydrates, proteins, and minerals. Most browsers are *concentrate selectors*, preferring parts with less cell wall and more cell content. They eat storage structures (seeds, fruits, roots), tissues active in metabolism (leaves, stems, flowers), or other plant products, such as nectar and gums or saps. The specific nutrients each of these food types offers depend on the species, the part consumed, the state of development or maturity, and other things. Ripe fruit flesh, for example, tends to be high in vitamins, simple carbohydrates, and water, whereas seeds and leaves are higher in protein and fatty acids. Roots and tubers tend to have more complex carbohydrates, water, and minerals, and nectars are almost all sugar and water.

Then there are faunivores. Within this category, many distinguish insectivores and carnivores, which specialize in small invertebrates and vertebrates respectively. Faunivores have some advantages and some disadvantages over herbivores. First, because animals tend to have similar nutrients, they are often easier to assimilate—no complex guts or large numbers of symbiotic microorganisms are needed. As with herbivores, though, faunivores have both benefits and costs to contend with. Animals don't contain much of the sugar needed to fuel the brain and some other tissues. So faunivores have to make glucose from the fat and protein of their prey—a process called *gluconeogenesis*. Also, because most animals do not want to be eaten, they defend themselves, hide, or move to avoid predators. Predation requires energy and can involve risk. Some animals are toxic or fight back. Many invertebrates have exoskeletons made largely of chitin, which is similar in structure to cellulose and so difficult to break into and digest. Also, because insects come in small packages, larger insectivores need a lot of them, so focus on colonial species. They develop special features such as long snouts, sticky tongues, and strong claws to deal with colonies—think of the aardvark, anteater, echidna, numbat, and pangolin.

Physical properties of foods. Animals may choose foods for their chemical nutrient properties, but teeth evolve with physical properties in mind. Dental functional morphologist Peter Lucas divides food properties into external ones (e.g. size, shape, stickiness, surface texture, abrasiveness) and internal ones (mechanical or fracture properties of a tissue). These properties are very important to think about when we consider the challenges teeth face. All of them, along with resistance to capture for animals, can affect food acquisition. They are also important for food processing, especially for those items that need to be broken before swallowing.

We use the terms *hardness* and *toughness* to describe how well foods resist the start and spread of a crack, respectively. Seed shells and bone tend to be hard, whereas mature leaves and skin may be tough. Dental researchers sometimes refer to these properties as defences, and distinguish them as *stress limited* and *displacement limited*. Stress-limited defences harden or stiffen an item to increase the stress, or force concentrated on a given area, required to start a crack. Displacement-limited defences divert or dissipate energy from the tip of crack to prevent its spread. Fracture properties are very important for teeth, because Nature should select the best tools available to break the types of food an animal has evolved to eat. You would not pair a knife with a walnut, or a nutcracker with a slab of raw meat. Nor would Nature.

Getting food in and out of the mouth

Given that different foods present different challenges, vertebrates have evolved different types of teeth to deal with them. But to understand how teeth work, we need to know more than how dental form relates to food fracture properties. We need to know how animals use their teeth—how they get food into the mouth, and how they prepare it for the gut.

Food acquisition. Getting food from the biosphere into the mouth can be a challenge. Potential foods may try to avoid being eaten by fleeing or defending themselves, they may be attached to something inedible, or they may simply be too big to fit into the mouth. Teeth can help a consumer overcome these challenges by capturing, holding, incapacitating, or killing prey, separating food items from non-food parts, and reducing them to bite-sized morsels.

So how does tooth shape relate to ingestive behaviour? Carnivorous fishes, amphibians, and reptiles often have conical or cylindrical teeth, *recurved*, or bent backward for engulfing and holding prey. The monk fish has a wide mouth with a long row of small but sharp mobile teeth that can incline inward but not out, to let prey enter but prevent escape. Even the seemingly innocuous salmon has long, scary-looking, needle-like teeth used often for puncturing and immobilizing prey. Teeth of some species are compressed side to side into knife-like structures with sharp edges, sometimes serrated, on the front and back to penetrate prey. This contrasts with the front teeth of sheepshead fish, which look eerily like human shovel-shaped incisors. Sheepsheads use these to grasp invertebrates, such as mollusks and crabs, and to scrape barnacles from rocks and pilings.

Most studies relating tooth form to ingestive behaviour have, however, focused on mammals. Some mammals take upward of 10,000 bites a day, so the pressure for efficient ingestion can be intense. Mammalian incisor size has been related to the amount of food an animal eats and how selective it is in choosing, as well as the degree to which these teeth are used in food acquisition, how they are used, and the forces acting on them during ingestion. For antelopes, incisor row length is a compromise between food intake rate and selectivity; high-volume grazers have broader incisors and muzzles than do fussier browsers. And for primates, those that regularly use their front teeth for husking large fruits have broader incisors than those that do not. Incisor size in rodents also relates to feeding rate. Faster is better, given both competition with neighbours and the need to limit exposure to predators that move between food patches looking for a quick meal. The ratio of incisor size to canine size can also be important: cats have bigger, stronger canines for deep, prolonged killing bites while holding struggling prey, whereas dogs have relatively larger incisors used to inflict shallow, slashing wounds and to gather other foods (see Figure 6).

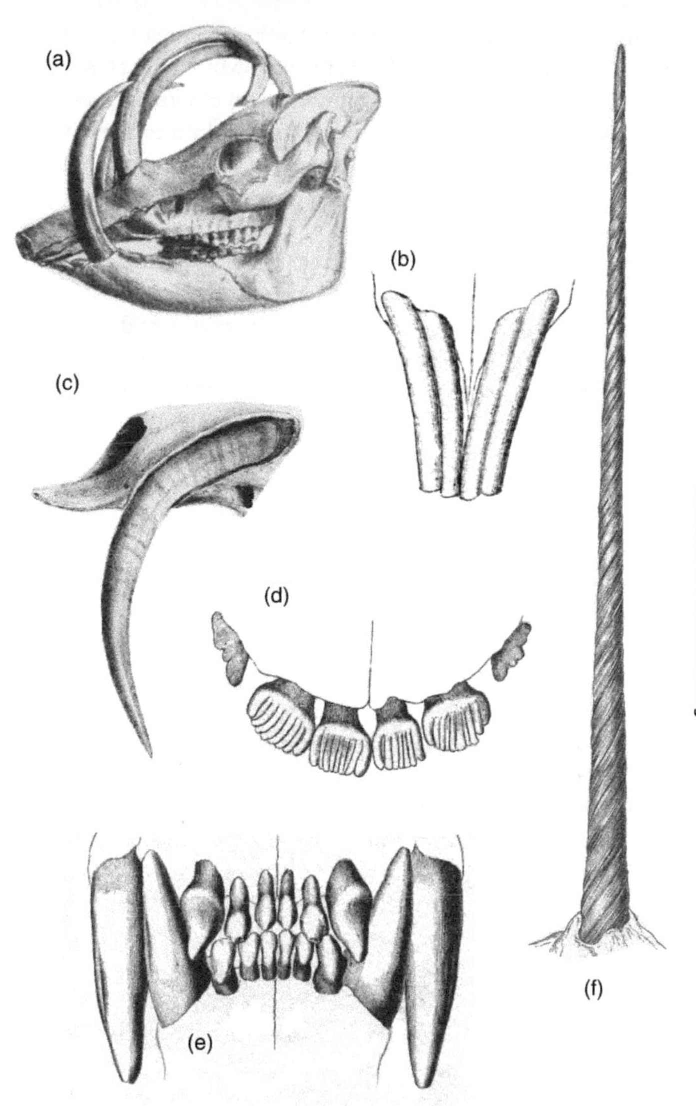

6. Mammalian front teeth. A, babirusa; B, cape mole rat; C, musk deer; D, colugo; E, lion; F, narwhal

There are other influences on front tooth size in mammals too. Some species, such as wombats and burrowing rodents, have large incisors to help them dig tunnels, and others, such as lemurs and tree shrews, have them modified into combs for grooming fur. And then there are tusks, which protrude over the lips when the mouth is closed. These enlarged incisors or canines are often used for display or fighting, but can also serve other functions. Elephants use them for digging and marking trees, and narwhals use them as sensory organs to detect changes in water temperature, pressure, and chemistry. Walruses even use their tusks as crutches to help them onto pack ice and rocks; their scientific name, *Odobenus rosmarus*, actually means 'tooth-walking sea horse'.

Food processing. Food processing means, for mammals at least, chewing. Teeth rupture protective casings such as plant cell walls and insect exoskeletons to access nutrients that would otherwise pass through the gut undigested. And breaking food into pieces decreases particle size for swallowing while increasing exposed surface area on which digestive enzymes can act. More surface area means more enzyme action, and more complete digestion. But chewing requires energy and takes time, so the costs must be weighed against the benefits. More time spent chewing means less time spent ingesting and less food eaten. In one recent study, men who increased the number of chews per bite from fifteen to forty decreased their calorie intake by 12 per cent. That's great if you want to lose weight, but Nature's goal is usually to maximize efficiency, not reduce it. And very small particles may pass through the gut too quickly, leaving too little time for bacteria to help with food breakdown. In this case, chewing too much actually decreases digestive efficiency. It's a balancing act. And more chewing can mean more tooth wear and less effective food breakdown, requiring even more chew cycles for a given morsel. It can become a vicious circle.

The fundamentals of mammalian chewing have been understood for a long time. The most important elements were described millennia ago by Aristotle in *De partibus animalium*. He wrote:

> Of the two separate portions which constitute the head, namely the upper part and the lower jaw, the latter in man and the viviparous quadrupeds [mammals] moves not only upwards and downwards, but also from side to side; while in fishes, and birds and oviparous quadrupeds [other vertebrates], the only movement is up and down. The reason is that this latter movement is the one required in biting and dividing food, while the lateral movements serve to reduce substances to a pulp. To such animals, therefore, as have grinder-teeth this lateral motion is of service; but to those animals that have no grinders it would be quite useless; and they are therefore invariably without it.

If it were not for some unfortunate basic errors, such as the claim that men have more teeth than do women, we might think Aristotle was clairvoyant. The three key points here are: 1) mammals and other vertebrates differ in how they chew, 2) the horizontal component to chewing is key to food breakdown for many mammals, and 3) chewing and tooth shape are matched for efficient food fracture.

Mammalian mastication is not unique because mammals chew but because of how they chew. Mammals are, as biomechanics researcher Callum Ross says, 'an extreme on the continuum'. They add a side-to-side, or sometimes a back-to-front, component to their jaw movements. We can call these transverse and longitudinal components respectively. Each chewing cycle is broken into three strokes: 1) recovery, when the mandible drops; 2) preparatory, when it closes and moves the lower teeth into position, so opposing occlusal surfaces approach from the correct angle; and 3) power, when forces are applied to food between upper and lower teeth as they come together and separate (see Figure 7).

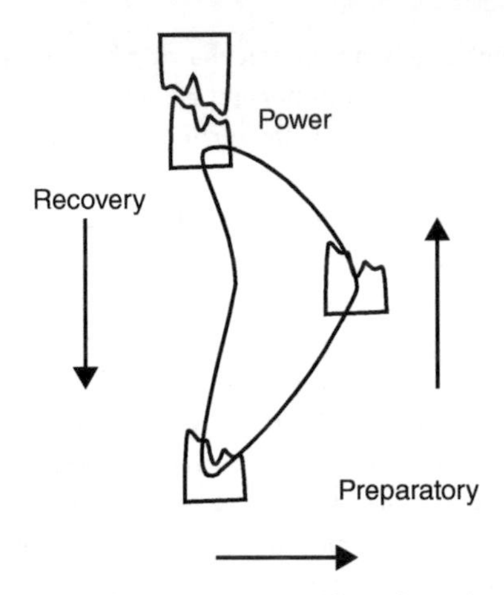

7. The chewing cycle of a sheep as viewed from front through molar teeth

How teeth break food

On one level, the shape of a cheek tooth can influence, or even guide, movements during the power phase of mastication. A lion can't grind its teeth because of the way opposing crests fit together during occlusion. On another level, the shape of the biting surface affects how it breaks food. A sharp blade is better than a blunt cusp for slicing tough meat. Georges Cuvier recognized the distinction between the two levels centuries ago in his descriptions of the teeth of hoofed animals, or ungulates. He noted that their teeth are flat to allow horizontal movements, but have uneven surfaces (alternating bands of enamel and dentine), to grind tough vegetation.

Tooth shape and diet. The great early 20th-century paleontologist George Gaylord Simpson envisioned teeth as guides for chewing. Some teeth have cusps that fit into opposing basins. These work well for crushing hard, brittle items, or for pulping fruits. The protocone of a typical primate fits neatly into an opposing talonid basin. Others have crests that slide past one another like scissor blades. This works well for slicing meat and other tough tissues. Cats and dogs have blade-like cheek teeth. Yet others have opposing surfaces with elements of both, for grinding. This works well for milling grass and other vegetation. Cows have crescent-shaped crests with alternating bands of enamel and dentine, and elephant molars have up to two dozen or more parallel ridges, each running side to side across the crown. In fact, you can get a pretty good idea of the diet of a mammal simply by measuring the relative shearing and crushing areas of its teeth. Bamboo-feeding pandas, shell-cracking sea otters, and nectar- and fruit-eating flying foxes have larger crushing and smaller shearing surfaces than do carnivorous polar bears, omnivorous badgers, and insectivorous serotine bats, respectively (see Figure 8).

But, as Peter Lucas has reminded us, studies of jaw movements alone cannot teach us how food is broken. Teeth tend to act by compression between the lowers and uppers. This is both obvious and counter-intuitive. Cracks should spread by pulling apart materials, not by pressing them together. Think of ripping a piece of paper. A relevant example is splitting a log. Pounding a wedge creates tension at the tip of the spreading crack. A less obvious example is cracking a walnut. When you crush it around the middle, it cracks at the ends, far from and perpendicular to the forces applied. Try it. These examples teach us how teeth fracture food.

Some researchers build models of idealized tooth shapes for foods with different fracture properties for comparison with real-world cases. The goal is to break without being broken. Stress-limited foods are often brittle; once a crack starts in one, it's difficult to stop it. On the other hand, it can take a lot of stress to start a crack in a hard food. A cusp tip is a good model because it concentrates force. If a cusp is too sharp, though, it can be easily damaged; a hemisphere should work well. We also need an opposing platform,

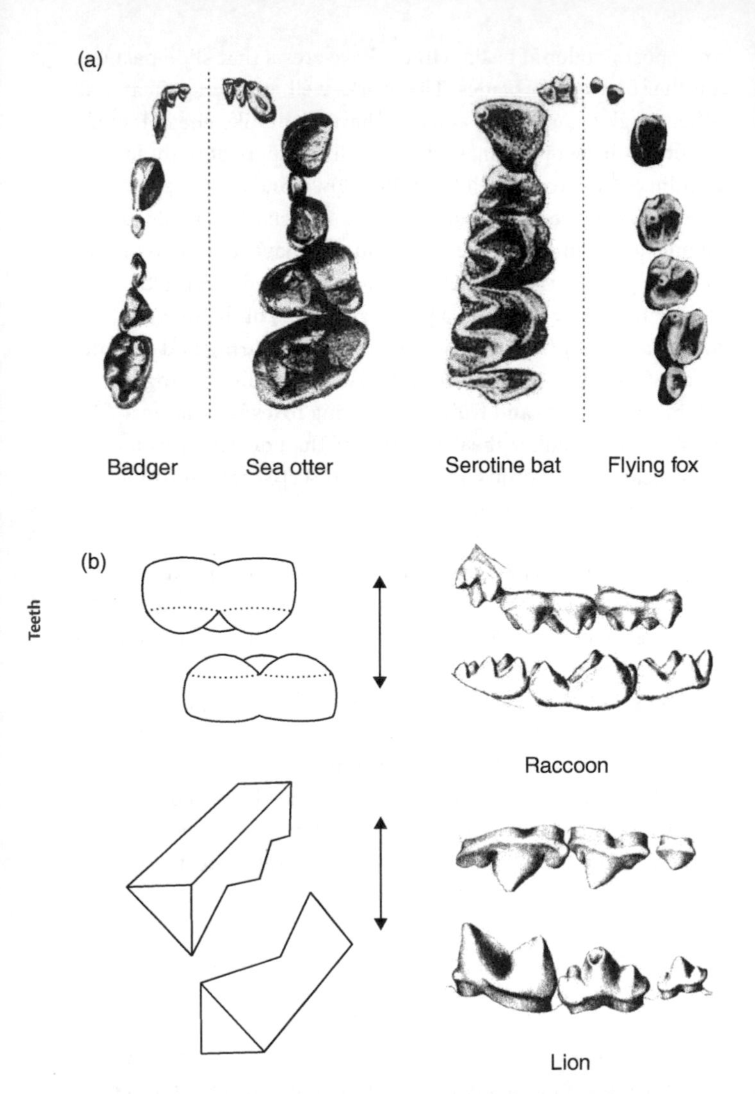

8. Tooth form and function. A, upper teeth of a badger and sea otter (left), and a serotine bat and flying fox (right); B, opposing idealized hemispheres modelling cusps and teeth of a raccoon; and opposing idealized blades modelling crests and teeth of a lion

which can be a basin or space between staggered opposing cusps. A concave surface, like a mortar, helps keep food in place. The molars of pigs and raccoons provide good examples. Teeth can also have fissures or troughs formed by alternating bands of enamel and dentine to direct food and fluids across the tooth and inward toward the tongue.

For displacement-limited foods, the challenge is less starting a crack than forcing it through the item. A sharp tooth is usually okay in this case, because the risk of damage is lessened as tough foods spread across its surface with compression. A wedge-shaped blade works well if it's thin enough to keep down energy needed to push the crack through but still thick enough to keep from breaking itself. The opposing surface could also be a blade, but the tips of uppers and lowers should be slightly offset so they slide past one another rather than colliding and damaging the teeth. Carnassials of cats and dogs are a great example. The last upper premolar and first lower molar have facing Λ- and V-shaped blades respectively, to trap tough animal tissue and prevent it from spreading out as it's sliced. And some blades have serrations, which can catch food and take advantage of its resistance and elasticity—think of a steak knife.

But relationships between tooth form and diet are complicated by wear. Teeth change shape over time. In many cases, the underlying structure of a tooth is actually laid out to guide wear and, in effect, sculpt the surface. Sharp edges form where thin layers of enamel are worn through to the underlying dentine. This is how cows and sheep get alternating bands of enamel and dentine, as Cuvier described. In fact, many rodents begin grinding their teeth in the womb, so they are sharpened and ready to go from birth. On the other hand, when tooth wear gets beyond a certain point, efficiency can begin to drop. Animals can chew longer, or eat more to compensate, but eventually starvation and death follow. Most mammals are in serious trouble when their teeth wear out. Some species have responded by increasing tooth height to extend life, especially those that eat gritty or abrasive foods that require a lot of chewing, such as grasses on the open savanna. We call teeth that are taller than they are long *hypsodont*. Others never stop growing, and keep adding tissue at the rate they wear, such as the gnawing incisors of rodents. These are called *hypselodont*.

Tooth size. The shape of a cheek tooth is important to how it fractures food, but size also matters. You might think that larger teeth mean bigger platforms for processing more food. If you have to eat more, say, because a given volume of whatever you are eating yields less energy, you should have bigger teeth. But it's more complicated than this because of the way tooth size varies with body size. An elephant has larger teeth than does a mouse just because it is bigger. The important question is: If you shrank an elephant to the size of a mouse, would their teeth be the same size? In fact, tooth size does not always vary one to one with body size across the mammals; and this has caused a bit of a stir among those of us that study teeth.

Early in the 20th century, Swiss biologist Max Kleiber observed that while larger mammals need more energy to power their bigger bodies, that need doesn't increase one to one with body size. For those mathematically inclined, Kleiber found that metabolic rate scales to the 3/4th power of an animal's mass. For those not, an elephant that weighs one hundred times your weight should burn about thirty-two times as much energy as you while at rest. At first glance, it makes sense, then, as paleontologists David Pilbeam and Steven J. Gould suggested, that tooth area should increase with body mass at about this rate. But, as another paleontologist, Richard Kay, has noted, larger mammals also tend to eat lower-quality foods. If you consider mammals with similar diets, tooth area increases with body size to the 2/3rd power, not the 3/4th power. Because tooth area is a two-dimensional measure, and body volume is three-dimensional, 2/3rd power actually means that mammals with similar diets increase their biting surfaces one to one with body size.

So what about Kleiber's rule? Smaller animals may have less efficient engines than do bigger ones. Because their surface areas are large relative to their volumes, they lose heat more quickly and need to burn more fuel, kilogram for kilogram. Still, as paleontologist Mikael Fortelius has noted, larger animals also chew more slowly than do smaller ones. He suggested that the difference in how quickly food reaches the gut matches the difference in needs between larger and smaller mammals—all else being equal. While this is all rather complicated, the take-home message is simple. If you enlarged a mouse to the size of an elephant, their teeth should be the same size if they are adapted to the same diet. On the other hand, lower-quality diets should select for larger back teeth given the need to process more food. And this actually does work in many cases, but not in all. Other variables, like space available in the jaw, sometimes come into play and muddle things up.

Foodprints. Researchers spend a lot of time relating tooth size and shape to diet in living animals, so they can infer diet from the teeth of fossil species. But there are other, more direct tools. We can call these foodprints. Like footprints in the sand, foodprints are traces of activities of individuals in the past. These include the chemical signatures that foods leave in teeth and patterns of tooth wear.

Foods differ in their chemical composition. They are made from different elements, and vary in proportions of variants, or isotopes, of specific ones, such as oxygen, nitrogen, and carbon. Because food provides the raw materials used to make bones and teeth, the chemistry of these tissues can give us important clues to diet. For example, plants have more strontium relative to calcium than do animals, and among plants, roots and stems have more than do leaves. So the ratio of strontium to calcium in a fossil can be used to reconstruct something about the diet of a long-extinct species if these elements are preserved unchanged from their concentrations in life.

What about isotopes? An oxygen atom has 8 protons in its nucleus, but can vary in number of neutrons, resulting in isotopes with slightly differing masses. We add the protons and neutrons to get the atomic mass number. If it has 8, we call it ^{16}O, if 10, ^{18}O. Water made from ^{16}O evaporates or transpires from leaves more quickly than does water made from the heavier ^{18}O. So animals that get water from leaves are ^{18}O enriched compared with those that get water from drinking, especially in dry environments. Researchers also look at how nitrogen isotope ratios change along the food chain. Carnivores have higher ratios of ^{15}N to ^{14}N than do their prey, and herbivores have higher ratios than do the plants they eat. The most common element used in diet reconstructions is carbon. Plants use energy from the sun to turn carbon dioxide and water into carbohydrate and oxygen in different ways. Most use what is called the C_3 photosynthetic pathway, which discriminates against CO_2 made with the heavy ^{13}C isotope. Most tropical grasses use a different pathway (C_4), which discriminates less against it. So, grazers in the tropics tend to have higher ratios of ^{13}C to ^{12}C than do browsers who eat trees, bushes, and shrubs.

The other commonly used foodprint is tooth wear. It comes in two varieties—mesowear and microwear. Mesowear begins with the notion that sharp facets form when opposing teeth rub together, but these abrade away with wear caused by food (or grit on it). And indeed, species of grazing hoofed mammals have blunted teeth compared with species that eat less-abrasive browse. Microwear is the study of microscopic scratches and pits on a tooth that form during use. Unsurprisingly, mammals that crush hard foods between opposing teeth tend to have pitted surfaces. In contrast, tough-food feeders tend to have more scratches, and these often run parallel to one another. This makes sense if abrasives on or in food are dragged between opposing blades

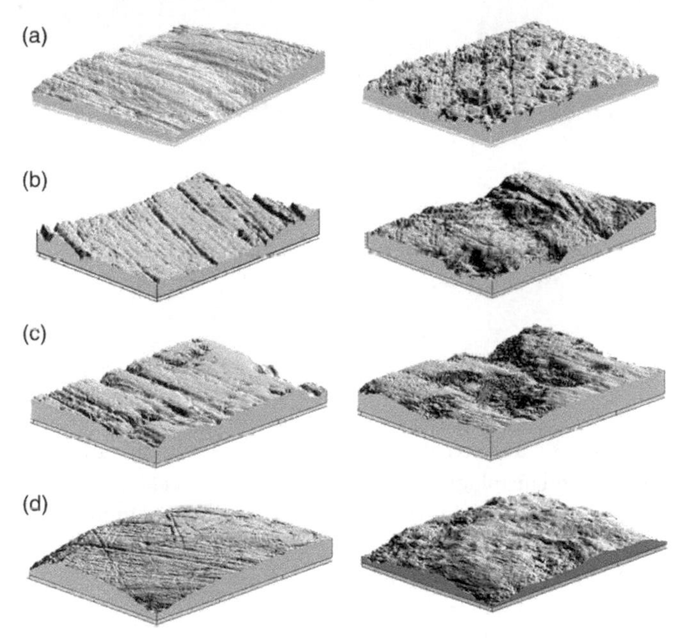

9. Dental microwear of A, grass- (left) and browse- (right) eating antelopes; B, leaf- (left) and nut- (right) eating primates; C, meat-eating cheetah (left) and bone-eating hyena (right); and D, carnivorous (left) and shell-crushing lizards (right). Each image represents an area 0.1 × 0.14 mm

as the lower teeth slide past the uppers. And differences in microwear reflect differences in diet for a very broad variety of vertebrates, especially mammals (see Figure 9). Nut-eating monkeys tend to have more microwear pits whereas leaf eaters have more scratches. Browsing antelopes, especially fruit eaters, have more pits, whereas grazers have more scratches. Bone-crushing hyenas have more pits whereas flesh-specialist cheetahs have more scratches. The examples go on and on.

Chapter 4
Teeth before the mammals

When did teeth first appear? Where did they come from? Researchers have used all the tools of the trade—comparative anatomy and histology, paleontology, embryology, and genetics—to address these questions. The answers are obscured in the haze of deep time, leading to controversy and bitter debate, but new insights are coming fast and furious.

Sea urchins, spiders, slugs, and squids

When you think about teeth, you probably envision sharks, dinosaurs, or even people. You think about jawed vertebrates, or *gnathostomes*. But many other animals, from slugs to spiders and sea urchins to squids, have similar structures in or around the mouth that function as teeth. They are not hardened with calcium phosphate like our teeth but, rather, calcium carbonate, chitin, or keratin. While these structures evolved separately from our teeth, they are important to consider for context. They serve as great independent examples of how Nature can meet the challenges of food acquisition and processing (see Figure 10).

Sea urchins have five of these structures that form part of a feeding apparatus called Aristotle's lantern. Aristotle described the apparatus in *Historia animalium* as resembling 'a horn lantern with the panes of horn left out'. The 'teeth' themselves are each

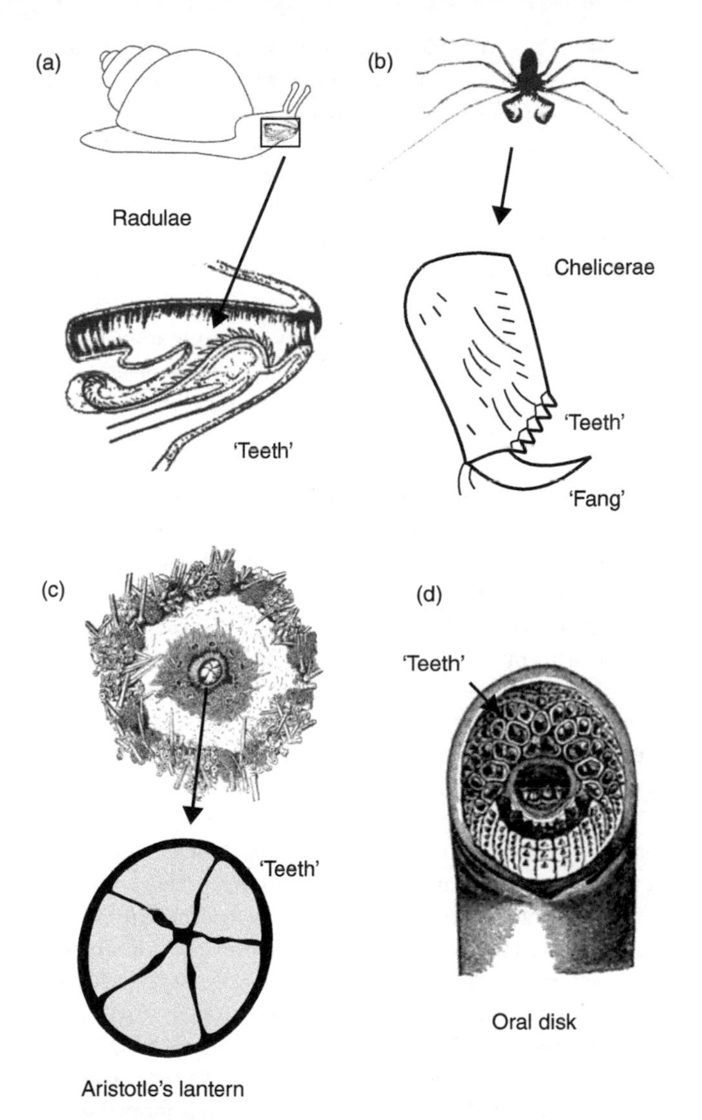

10. 'Teeth' of invertebrates and the lamprey. A, radular 'teeth' of a mollusk; B, chelicerae of a spider; C, Aristotle's lantern of a sea urchin; D, oral disc and 'teeth' of a lamprey

shaped like curved triangles that, when brought together, form a dome. The dome points outward from the mouth and opens and closes with tips of the 'teeth' moving in and out of contact to rasp algae from rocks. These also function to bore into limestone to create hollows for shelter. Remarkably, these 'teeth' remain sharp, even though they and the rock are both made of calcium carbonate. How? Sea urchin 'teeth' grow continuously, and have a microscopic structure laid out with planes of weakness leading to a predetermined breakage pattern that maintains sharp edges. They are also hardened with magnesium, especially at the tips. Sea urchins have recently caught the attention of engineers interested in using biologically inspired designs to create self-sharpening tools.

Spider *chelicerae* also come to mind. The basic model is a fang attached to a base by a hinge joint. The fang pivots into a groove on the base, like a pocketknife being opened and closed, to crush food items beneath it. Oftentimes the edges of the groove are lined with pointed 'teeth', typically up to about fifteen in one or two rows. These 'teeth' vary in size and shape between species, and even within rows. The inner surface of the fang is also sometimes serrated, like a steak knife, to help shred food. This extraordinary structure serves both to deliver venom and to mash prey into a soft pulp for swallowing.

Then there are the mollusks. Tens of thousands of species, from slug to snail to squid, have 'teeth'. These form in rows on ribbons of chitin in the mouth called *radulae*. Many mollusks use these structures as a comb to rake up microorganisms, or as a rasp to scrape food from rock or shell. Radulae typically move back and forth like a handsaw. While radular 'teeth' tend to be small and recurved, shapes and sizes can vary with species and function in feeding. They can even vary within individuals. In fact, a change in diet can trigger a change in shape for new 'teeth' formed to replace old, worn ones. Also, some radulae are extremely specialized. Whelks, for example, commonly have three long, sabre-like 'teeth' in each row. These are used to drill through barnacle and clam shells with the help of secretions that break down calcium carbonate. And cone snails have radulae modified into hypodermic needles to inject venom. These have barb at their ends, and can be extended from the mouth like harpoons to attack and paralyse prey.

Finally, there are the hagfishes and lampreys. These are jawless fishes whose ancestors split from the line leading to the gnathostomes before jaws evolved. Like sea urchins, spiders, and slugs, their 'teeth' have little to do with ours. They are made of keratin, like your fingernails. Hagfishes have two rows of sharp, recurved 'teeth' on each of two dental plates that unfold as they extend from the mouth. Items are hooked on the 'teeth' as the plates retract and fold back in toward one another. Hagfishes typically use them to rasp flesh from dead or dying animals. Lampreys, on the other hand, at least most adult marine forms, are parasites, and use sharp 'teeth' lining their oral discs to pierce and latch on to live prey with help from a suction-cup-like mouth. They also have a tongue-like structure used to rasp skin, and secrete an anticoagulant to keep the blood of their host flowing.

The origin of *real* teeth

But what about real teeth? Goldfish have them. People have them. Teeth were passed along to each of us from a common ancestor; in biology speak, they are *homologous* structures. Imagine the very first animal on our evolutionary line to have them. We consider teeth 'real' if their inheritance can be traced to that distant ancestor.

What do we know about this ancestor? We know it was a vertebrate. Recall from the discussion on dental development that the embryos of complex animals divide into three layers, the endoderm, mesoderm, and ectoderm, early in development. Also recall that vertebrates have a fourth layer, the neural crest, which comes from the ectoderm. Neural crest cells differentiate and migrate to different parts of the embryo. These interact with other cells to form various structures, including the teeth. Because it takes a neural crest to make a tooth, we can limit our search for real teeth to the vertebrates.

Outside in. Our first clue comes from sharks. Sharks and rays have skin covered in small, tooth-like structures called *placoid scales.* These give shark skin its rough, sandpaper-like texture. Placoid scales are tiny cones of dentine with a cartilage base and an internal pulp cavity housing blood vessels. Sound like teeth? The resemblance has led to the popular theory that teeth are modified scales that migrated to the mouth from surrounding skin when the jaw evolved—from outside in. Placoid scales make good models for the precursors of teeth, and both are variants of a fundamental unit with the same basic structure, called an *odontode.*

If the outside-in hypothesis is right, we should find tooth-like scales before true teeth in the fossil record. Do we? The first known vertebrates date to early in the Cambrian, at least 530 million years ago (mya) (see Table 1). These aren't much help to us because they have neither scales nor teeth. A group of slightly younger vertebrates, the ostracoderms, may offer some clues, though. These jawless fishes appeared later in the Cambrian, about 500 mya, and dominated the seas for nearly 100 million years. Ostracoderms had a scaly tail and head armour made from tiny hardened plates of calcium phosphate. Each plate had an outer surface of dentine, sometimes capped with a more mineralized, enamel-like tissue, all covering a pulp chamber that housed blood vessels. These plates also had underlying layers of spongy and lamellar bone. Ostracoderms did not have teeth, but some had odontode-like plates on the rim of their mouth that probably functioned in feeding, with small nubs or barbs likely used to sieve microorganisms from the water around them.

Table 1. Geologic timescale for the Phanerozoic Eon

Era	Period	Start (mya)
Paleozoic	Cambrian	541.0 ± 1.0
	Ordovician	485.4 ± 1.9
	Silurian	443.4 ± 1.5
	Devonian	419.2 ± 3.2
	Carboniferous	358.9 ± 0.4
	Permian	298.9 ± 0.15
Mesozoic	Triassic	252.17 ± 0.06
	Jurassic	201.3 ± 0.2
	Cretaceous	~145.0
Cenozoic	Paleogene	66.0
	Neogene	23.03
	Quaternary	2.588

Based on the *International Chronostratigraphic Chart v 2013/01* (International Commission on Stratigraphy)

Inside out. On the other hand, there are the conodonts, a diverse group of eel-like animals that lived between at least 510 and 220 mya. Conodonts lacked the hardened scales and dermal armour of the ostracoderms, but had tiny tooth-like elements made from calcium phosphate and assembled into sets within the head, around the area of the throat. These elements come in a remarkable range of shapes and sizes, from simple, cone-like forms to complex arrays of elaborate 3D structures. Studies of their morphology and wear suggest that conodont elements were used to shear and grind food hundreds of millions of years before our earliest mammalian ancestor first brought its upper and lower teeth together. In fact, conodonts may have been the first animals to experiment with chewing. This at first glance suggests that teeth started in the throat and moved out to the edge of the mouth later—the inside-out hypothesis (see Figure 11). If so, teeth evolved before jaws, which may sound surprising to you and me, but not to a zebrafish, which has teeth in its pharynx but not its mouth. In fact, many fishes have pharyngeal teeth. These can be quite elaborate, opposing one another or hardened plates to crush food as it passes through the throat. Further, placoid scales, pharyngeal denticles, and teeth are all formed by the same set of genetic controls: they are *serial homologues*. Evolution commonly works through serial homology, replicating existing parts and modifying or building on to them to meet new needs. Think of your arms and legs.

That said, the microscopic structures of conodont elements and gnathostome teeth are actually quite different, and there is no evidence of evolution from one to the other. It is actually more likely that conodont elements and gnathostome teeth developed independently.

This puts *Loganellia scotica* of the Silurian (nearly 440 mya) at centre stage. *Loganellia scotica* was more clearly a vertebrate and, while it lacked oral teeth and jaws, it had both dermal scales and pharyngeal denticles. Its denticles are joined into sets that look more like teeth than do its scales, perhaps offering better evidence for the inside-out hypothesis. But its pharyngeal denticles may not have been homologous with teeth either. While they have the same general odontode structure as teeth, they were laid out differently, and many of *Loganellia*'s close relatives did not have these structures, so they also may have evolved independently in this species, separate from ancestral gnathostome teeth. If one group can evolve tooth-like structures, why not two groups, or three, or more? We call this *homoplasy*, and it wreaks havoc when we are trying to figure out how fossil species are related, and where, when, and from whom anatomical features evolved.

Early jawed fishes. Most jawed vertebrates today have teeth (see Figure 12). Perhaps, then, the best place to look for evidence of

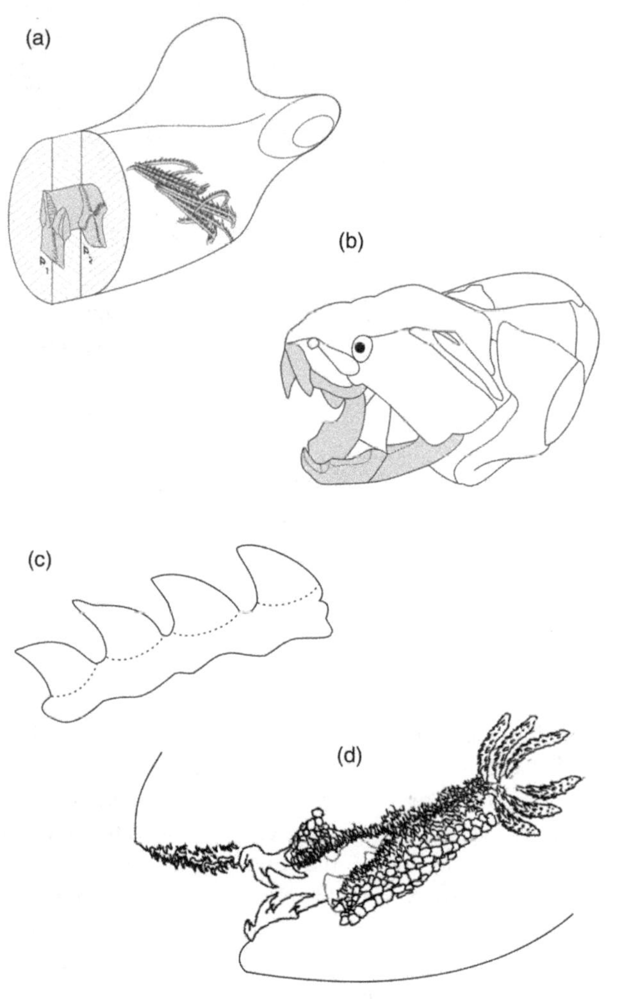

11. Dental structures in Paleozoic fishes. A, conodont elements (*Idiognathodus*); B, placoderm oral plates (*Dunkelosteus*); C, thelodont pharyngeal denticles (*Loganellia*); D, acanthodian scales and teeth (ischnacanthid)

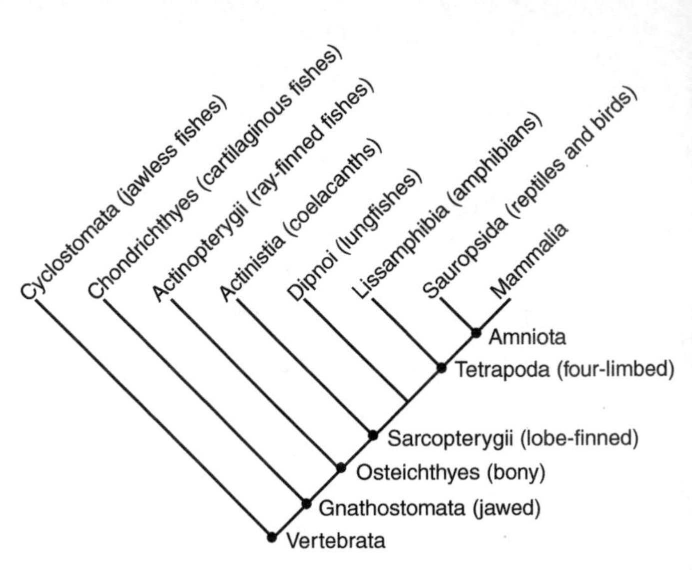

12. Relationships among the vertebrates

the origin of teeth is the earliest gnathosomes. These have traditionally been divided into the wholly extinct acanthodian and placoderm fishes on the one hand, and the chondrichthyans (cartilaginous fishes, or sharks, rays, and chimeras) and osteichthyans (bony fishes and land vertebrates) on the other. These are likely not all natural groups. Some acanthodians were probably more closely related to bony fishes than to other acanthodians, and some placoderms were evidently more closely related to the living fishes than to other placoderms. But until their relationships are better worked out, this grouping at least gives us a convenient way to structure our search for the earliest teeth.

We can start in the oceans of the Silurian Period (see Table 1 for a timescale). Evolution was in overdrive. Not only did the jawless vertebrates diversify during the period, but we also get the earliest undisputed evidence of all four gnathostome groups—first the acanthodians and placoderms, then the osteichthyans and chondrichthyans. The acanthodians resembled, but were not, small sharks. They appeared early in the Silurian, if not before, spread into a variety of aquatic ecosystems during the period that followed, the Devonian, and survived 150 million years, all the way to the Permian. Most early acanthodians had upper and lower jaws but no teeth. They had instead tiny, finger-like spines called *rakers* to filter food suspended in water before it entered the gills. But some had teeth. Some of those had tooth whorls lining the jaw, basically spiral or arched cog-like conveyor belts, with sharp, recurved cones or triangles rotating into place for use. Others had rows of individual teeth fused to the jawbone, added one by one to the front, with those behind becoming worn or broken. Yet others had both kinds of teeth.

Acanthodians teach us a lot about the origin of teeth. Unlike most fishes today, they did not shed and replace their teeth, and they had no enamel or enameloid covering the dentine. This suggests that early gnathostomes figured out how to make teeth before learning out how to replace them or strengthen them with a highly mineralized cover. Also, some researchers have argued that acanthodians are more closely related to bony fishes than to sharks. If so, replacement and hardened tooth caps evolved independently in the osteichthyans and chondrichthyans. Finally, some acanthodians seem to show a transition between head scales and teeth. The ischnacanthids, for example, have lip and especially cheek scales that look like tooth whorls, increasing in size with proximity to the mouth. And like tooth whorls, but not the pharyngeal denticles in *Loganellia scotica*, the cusps get larger as new ones are added. This may well be evidence for the outside-in hypothesis.

Then there are the placoderms. These jawed fishes also appeared early in the Silurian and flourished in the Devonian. Placoderms dominated marine and freshwater environments during their heyday, but they apparently did not survive the ecological crisis and mass extinctions at the end of the Devonian. The placoderm head and thorax were shielded with thick, bony armour, and early species had dental plates covered by small spikes or denticles. Researchers debate whether to call these teeth, especially the more tooth-like structures in the advanced placoderms called arthrodires. Cusps were added in succession, and made of dentine with an underlying pulp cavity that was filled with tissue during life. Like the acanthodians, arthrodires did not shed or replace these structures. If primitive placoderms lacked teeth but arthrodires had them, teeth probably evolved separately in this group and modern fishes, unless the primitive placoderms and arthrodires each have separate ancestries. Whether these are real teeth or not, however, no late Devonian fish with any sense of self-preservation would have ignored the largest arthrodires. The giant 10-metre long predator, *Dunkleosteus*, for example, had long, razor-sharp dental plates that would have made the fiercest living great white shark turn tail and swim for its life.

Milestones and trends

Once teeth were in place, the focus shifted to making them work better. Important milestones and trends become clear when we compare sharks and other cartilaginous fishes to bony fishes, bony fishes to amphibians, amphibians to reptiles, and reptiles to mammals. Enamel evolved, along with new ways of attaching tooth to jaw. There were trends toward reduction in the number, distribution, and replacement of teeth. And while we usually think of mammals when considering different kinds of teeth in one mouth, complex crowns, occlusion, and chewing, many other vertebrates experimented with one or more of these things. After all, teeth have been around twice as long as mammals (see Figure 13).

Enamel. Vertebrate teeth are commonly covered by a highly mineralized cap, enameloid in most fishes and enamel in most

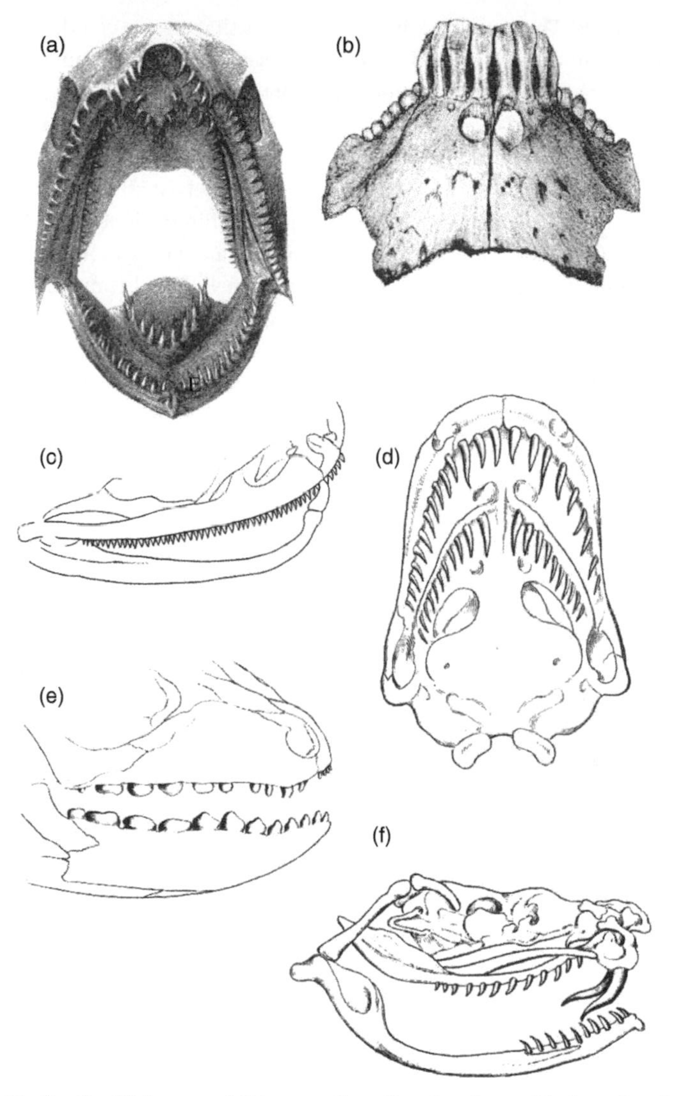

13. Teeth of fishes, amphibians, and reptiles. A, salmon; B, sheepshead fish; C, frog; D, caecilian amphibian; E, flying dragon lizard; F, pit viper snake

tetrapods. These are both hardened tissues that strengthen the tooth; but they develop differently. While enamel forms from ameloblasts, enameloid requires the combined action of ameloblasts and odontoblasts. This has important implications for the underlying structure and the chemical composition of the tissues. Genetic studies suggest that enamel evolved after the split of cartilaginous and bony fishes, probably within the lobe-finned fishes that gave rise to the tetrapods more than 350 mya. But the complex prisms typical of mammalian enamel did not come until much later, in the Mesozoic.

Tooth attachments. Different species have their teeth attached to the jaw in different ways. They can be fixed to the tip or the side, or they can be embedded in sockets. They can be connected by bone or by a periodontal ligament. And they can be attached individually or in groups by a common tissue. Cartilaginous fishes typically use a common sheet of connective tissue, and bony fishes attach them individually. Bony fishes tend to have their teeth attached to the tip of the jaw, whereas amphibians and most reptiles have them connected to the side. Only a few fishes, crocodiles, and mammals have tooth sockets today, though many more, such as toothed birds and dinosaurs, had them in the past. In crocodiles, successive teeth erupt in the same sockets as their predecessors, but in mammals, the walls of sockets are replaced by new bone once the milk teeth are shed and adult ones erupt. Also, crocodile teeth are fixed in the socket by a partially mineralized ligament. This condition is intermediate between the mammalian periodontal ligament and the more primitive vertebrate bony attachment.

Reduced number, distribution, and replacements of teeth. Fishes can have thousands of teeth in the mouth at one time. Amphibians have fewer teeth, but typically still have more than reptiles. Mammals tend to have fewer still. There are exceptions to this trend, though, and some species within each group have fewer teeth than others, or have lost them completely.

The placement of teeth in the oral cavity also varies among groups. Fishes tend to have their teeth spread throughout the mouth and throat, whereas amphibians and reptiles have more restricted tooth distributions, though they are still often attached to several bones of the skull. Mammalian teeth are confined to the margins of the mouth, implanted in only two or three bones. In our case, they're the maxilla and mandible.

Finally, there is something of a trend toward reduced number of tooth replacements. Sharks can shed and grow new teeth 200 times, whereas crocodiles have about forty-five to fifty generations. Mammals replace their teeth only once or not at all. Some other vertebrates, such as lizards (living and fossil) that have converged with mammals on the need for precise occlusion, also have reduced numbers of replacements.

Crown differentiation. We usually think of fishes, amphibians, and reptiles as having simple, peg-like teeth. But when you combine hundreds of millions of years with evolvable structures such as teeth and a drive to nourish the body, nature can do better than simple pegs. And it often does. There are about 28,000 species of fishes, living in almost every imaginable watery habitat and eating foods ranging from some of the smallest to the largest organisms on the planet. Of course their teeth vary. The horn shark, for example, has sharp, pointed front teeth for securing prey, and thick, rounded ones behind for crushing sea urchins and hard-shelled mollusks and crustaceans. Fish teeth range from widely spaced and spike-like to closely packed and multi-cusped. Some even have ridges and crests on their surfaces. And not only do teeth vary by position in the mouth (or throat for that matter), but replacement teeth often differ from preceding generations.

Amphibian teeth are less variable. Most have a distinctive ring of dentine or fibrous connective tissue separating the root from the crown, which makes it look like the tooth is on a pedestal. In fact, these teeth are called *pedicellate*. Their crowns are often simple pegs, but sometimes have two cusps. And fossil species add more variation to the assortment of amphibian tooth forms. The late Paleozoic lepospondyls, a diverse group of small newt- and eel-like species, tended to have more bulbous tooth crowns with modest cusp development. And the more recent (mid Jurassic to early Neogene), salamander-like albanerpetontids had reasonably complex multi-cusped teeth, at least by the relatively undemanding standards of amphibian dental morphology. Amphibian tooth replacements can also differ from their predecessors. Consider the caecilians, legless amphibians that look like worms or snakes. Adults can have dozens of small, sharp teeth, which are impressive in their own right; but hatchling teeth look like little grappling hooks, and are used to peel and eat their own mother's skin.

Moving on to reptiles, the lepidosaurs (lizards and snakes) often have pointy, recurved front teeth but more complex back ones. The iguana has blade-like back teeth, compressed side to side, for shredding vegetation. These have several cusps or large serrations that give each tooth a leaf-shaped appearance. The Komodo dragon, in contrast, has long and sharp recurved teeth with fine serrations, especially on the back end, for slicing flesh. The closely related Gray's monitor lizard has blunter, rounded crowns for crushing fruit and snail shell. And some whiptails have two or three cusps side by side on molar-like back teeth. Fossils again add variation to the mix. *Polyglyphanodon*, from the late Cretaceous, had molar-like marginal teeth with inner and outer cusps connected by a sharp, V-shaped blade with tiny serrations along its tip, like a cross between a dog or cat carnassial turned sideways and a steak knife. And like those of fishes and amphibians, reptilian replacement teeth may differ from their predecessors, so form can change with age.

The archosaurs (crocodiles, birds, and their kin), in contrast, use a different strategy for food processing. Living crocodiles have simple cone-shaped teeth, and birds have none. The latter grind their food with small stones, or gastroliths, housed in a muscular stomach chamber called the *gizzard*. Bird gizzards are lined with keratin, and vary in shape and muscularity depending on food properties. This works well enough that birds and mammals have about the same digestive efficiency. It also reminds us that while teeth are the mammalian solution to the problem of heating an energy-hungry, endothermic body, they are not the only solution.

Still, fossil archosaurs often had much more elaborate tooth crowns than their living descendants. The extinct crocodile *Chimaerasuchus* from the early Cretaceous of China, for example, had molar-like upper teeth with three rows of seven recurved cusps, each running front to back. And the dinosaurs had incredible variation in tooth form (see Figure 14). We can start with the saurischians, or lizard-hipped dinosaurs. These include the theropods (mostly bipedal carnivores, such as *Tyrannosaurus rex*) and sauropods (large, long-necked herbivores such as *Apatosaurus*, which used to be called '*Brontosaurus*'). Theropods frequently had recurved and flattened dagger-like teeth lined with sharp, serrated edges or tiny hook-like projections, presumably for grasping prey and ripping flesh. Sauropods in contrast, often had rows of tiny peg- or cone-shaped teeth, presumably for cropping vegetation.

But it was with the ornithischians, or 'bird-hipped' dinosaurs, that archosaur dental form reached its pinnacle. Many of these dinosaurs had extremely ornate tooth crowns and differences in shape between front and back teeth suggesting a dental division of labour. The heterodontosaurids, for example, had small, peg-like front teeth, enlarged, canine-like tusks, and complex marginal teeth, often chisel shaped with ridges or denticles along the biting edge. Many ornithischians had lance-shaped cheek teeth with serrations or denticles on blades forming the front and back edges. Enamel is often thicker on the outer side of the uppers and inner side of the lowers, and there may be none at all on the opposite

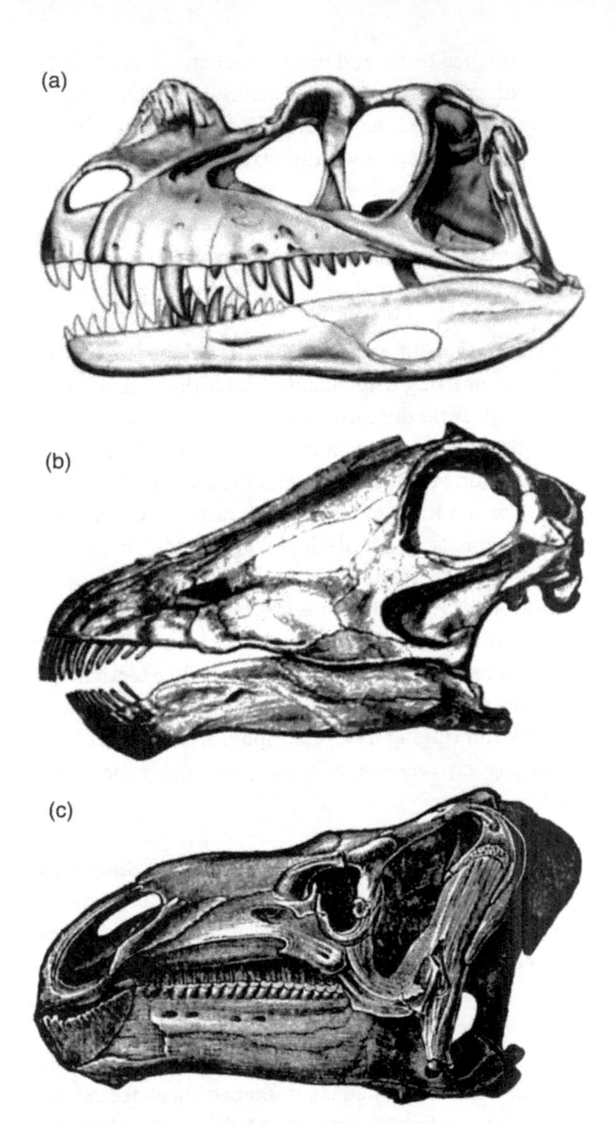

14. Dinosaur skulls and teeth. A, theropod (*Ceratosaurus*); B, sauropod (*Diplodocus*); C, hadrosauriform (*Iguanodon*)

sides. Since dentine is softer than enamel, the edges became sharp with wear as opposing teeth scraped across one another during feeding. But no other dinosaurs had teeth as elaborate as those of the hadrosaurs and ceratopsids. The duckbills, and horned, frilled dinosaurs had a unique arrangement of dental batteries, with dozens of individual teeth packed tightly and interlocked to form long, often continuous surfaces. Successive generations of teeth are also interlocked in the jaw, one above another, and with those of adjacent tooth families. The result is a 3D array that would have allowed efficient, uninterrupted shredding and milling of tough vegetation throughout life.

Occlusion and chewing. Imagine trying to cut using scissors with blades that don't line up. Chewing demands precise alignment. Cusps must fit into basins, crests against opposing crests, etc. Not only do upper and lower teeth have to match, but the chewer must have sufficient muscle control for fine-scale movements of the jaw to bring opposing surfaces together, and a joint that allows those movements. Precise occlusion and chewing in land vertebrates goes back about 300 million years, to a group of large, reptile-like tetrapods called the diadectids. This family was diverse and widespread in the late Paleozoic. Diadectid marginal teeth were broad and bulbous, especially in adults, and opposing surfaces have matching areas of wear that provide telltale clues to how lower and upper teeth came together during feeding. Microscopic scratches on those wear facets run back to front, indicating that the lower jaw and attached teeth slid forward, rather than side to side, during chewing. This longitudinal movement is called *propaliny*. The tuatara, a lizard-like reptile from New Zealand, and many rodents today use this type of movement to mill or grind their food.

The duckbill dinosaurs and their kin developed an even more ingenious and sophisticated chewing system. Like most living reptiles, these dinosaurs had simple, hinge-like jaws that allowed only vertical opening and closing. But like mammals, they had vertical and side-to-side movements between opposing occlusal surfaces. How could they accomplish this? The lower cheek teeth were bevelled with the inside edge higher than the outside, and the uppers were the opposite. When the lower jaw was raised, its teeth acted as a wedge, forcing the left and right upper rows apart (the two halves of the upper jaw were not fused). When the mouth was open, muscles or ligaments pulled the upper left and right jaw halves together, rotating the upper teeth back inward. But if you think this is complicated, just wait for the mammals.

Chapter 5
The evolution of teeth in mammals

Think about what happens in your mouth the next time you eat something. Your jaw, throat, and cheek muscles, tongue, teeth, and salivary glands all act in concert with sensory feedback to capture, transport, chew, and swallow food. The alignment and movements of opposing teeth are precise to a fraction of a millimetre as you generate, direct, and dissipate the forces needed to break food. You position and hold objects in your mouth and keep air and food passages separate to prevent choking. All this is carefully coordinated, with the various parts working together in symphony and synergy. How could this incredible system have evolved? The answer is written in stone—documented in a fossil record spanning hundreds of millions of years. This record is the story of the origin and evolution of the mammals.

Three great waves

We begin late in the Carboniferous, around 310 mya. The early amniotes, ancestors of the reptiles, birds, and mammals, have evolved an egg that can be laid, incubated, and hatched on dry land. They are freed from the shackles of an aquatic environment, and species adapted to the new opportunities and challenges of a fully terrestrial lifestyle are beginning to proliferate. Three types of amniote evolve, distinguished by the number of holes, or 'windows', in the sides of their skulls: 1) anapsids with no holes,

2) synapsids with one, and 3) diapsids with two. All living reptiles are diapsids (even the turtles, which have anapsid-like skulls today); and the mammals are synapsids. But the first synapsids came well before the mammals. In fact, they were among the earliest of the amniotes. Synapsids evolved in three great waves: first the pelycosaurs, then the therapsids, and finally, the mammals. Each wave developed into an extraordinary radiation of species, and each of these radiations became the dominant land vertebrates of their time.

The pelycosaurs. The pelycosaurs ruled the warm, wet equatorial ecosystems of the supercontinent Pangea during the late Carboniferous and early Permian. Some were carnivores, with sharp, cone-shaped teeth, including a couple of large, canine-like ones in each quadrant of the mouth. Others were herbivores, with blunt and sometimes leaf-shaped teeth compressed side to side and with coarse serrations on the front and back edges. Pelycosaurs came in many different shapes and sizes, but the sail-backed *Edaphosaurus* and *Dimetrodon* are both favourites in natural history museum displays the world over. *Edaphosaurus* had small, peg-like teeth rimming its jaws, presumably for cropping and grinding tough vegetation. *Dimetrodon*, in contrast, was a top predator of its day, with enlarged incisor-like and especially canine-like teeth, but small, sharp marginal teeth shaped like recurved steak knives, with serrated cutting edges on the front and back surfaces. The pelycosaurs were very successful during their heyday and thrived for tens of millions of years, but they ultimately declined as atmospheric carbon dioxide levels, global temperatures, and seasonal aridity increased. They were gone before the end of the Permian.

The therapsids. The therapsids arose from within the pelycosaur radiation and ultimately replaced it. These reptiles flourished under the changing conditions of the middle part of the Permian and at higher latitudes, likely because they were better able to control their body temperature and water balance. These were even more mammal-like, with upright limbs pulled under the body, a higher metabolic rate, and perhaps even hair and lactation. While researchers debate when they first appeared, a diverse community was in place by about 265 mya; and the therapsids thrived for the rest of the Permian. They ranged from a few centimetres to six metres long, from specialized burrowers to swimmers, and from bulk-feeding herbivores to top predators. And many had rather complex teeth. Some species developed interlocking, opposing incisors; and long, sabre-like canine tusks were common. The well-known dicynodonts often had a pair of tusks on each side of the upper jaw—hence the name, meaning 'two dog teeth'.

Then, around 251 mya, came the end-Permian event, or events. Paleontologist Michael Benton calls this 'the greatest mass extinction of all time'. It may have been triggered by impact from a comet or asteroid, by massive volcanic eruptions, or perhaps the unfortunate coincidence of both. Up to 96 per cent of all species on the planet were lost in a geological heartbeat. And it took fifteen million years for global ecosystems to recover. Only a handful of therapsid species survived, but their descendants emerged as important predators and herbivores when the dust settled in the post-apocalyptic Triassic. Some of the herbivores had pretty sophisticated cheek teeth with rows of cusps and crests that fitted between one another in an alternating fashion for milling tough vegetation using back-to-front chewing motions. Still, the mammal-like reptiles never again dominated the landscape as they did in the late Paleozoic. By the late Triassic, the archosauromorphs, especially the dinosaurs, began to radiate and take over terrestrial ecosytems.

The earliest mammals. But the synapsids did not disappear. The earliest mammals emerged from within the therapsid radiation in the late Triassic. Unlike the pelycosaurs and therapsids that preceded them, though, the mammals did not come to prominence quickly. Think of the classic museum diorama with a small, nocturnal insect eater lying meekly in wait for the rock to drop and end the reign of the dinosaurs. It was not until the Cenozoic that mammals really began to dominate the landscape; but they did radiate modestly during the Mesozoic, and this set the stage for things to come. We can envision early mammals eking out a living during the cold, dark night, with adaptations to improve temperature regulation and thermal insulation. It takes a lot of fuel to heat the body, and pressures on the teeth for more efficient food acquisition and processing must have been intense. As hearing and smell became more important, energy-hungry olfactory and auditory lobes of the brain expanded. As paleontologist Tom Kemp has suggested, this may have touched off a feedback loop of increasing feeding efficiency and enhanced senses to navigate an increasingly challenging food web. Mammals, it is said, were the end result.

The keys to mammalian chewing

There are about half a dozen things we look for in the fossil teeth and skulls of synapsids for evidence of the evolution of mammalian chewing, or mastication. These include separation of the front and back teeth into different types, a new jaw joint, reorganization of the chewing muscles, two generations of teeth, a bony palate, and prismatic tooth enamel.

Dental division of labour. While some fishes, amphibians, and reptiles have different kinds of teeth in different parts of the mouth, mammals, as already noted, take the dental division of labour to a new level. There are hints among the pelycosaurs. The name *Dimetrodon*, for example, comes from the Greek *dimetros* (two measures) and *odon* (tooth). And this is actually an understatement. These mammal-like reptiles had *three* types of teeth: thick front teeth, long, sharp canine-like teeth, and recurved, blade-like back teeth. By the time we get to the therapsids, we can reasonably start calling the different tooth types incisors, canines, and postcanines. The cynodonts, or 'dog-tooth' therapsids, were especially mammal-like. Many had increasing crown complexity from the canines to the back of the row, foreshadowing the separation of premolars and molars. And some had elaborate cheek teeth, from bladed structures for slicing to those with many rows of small, crescent-shaped cusps for grinding. By the late Triassic, one group of cynodonts, the trithelodontids, evolved teeth like those we'd expect of the ancestral mammal. Trithelodontid cheek teeth have a single row of cusps running front to back and connected by crests. The locations of use-wear facets on those teeth indicate that they functioned as shears for slicing food items.

Reorganization of the chewing muscles (see Figure 15). The earliest amniotes had simple hinge-like jaw joints. Their lower jaws were raised using muscles called *adductors*, which formed a sling around the mandible. An internal group connected the palate to the inside of the lower jaw, and an external one ran from the side of the cranium to the outside of the mandible. This worked well enough for closing the mouth, but not for the precise side-to-side, or backward-to-forward, movements needed for mammalian mastication. Mammals need finer control of jaw movements.

Mammals today still have a single inner part to the sling, the *medial pterygoid*, but the outer part has separated into two distinct muscles, the *temporalis* and the *masseter*. You can feel your temporalis flex when you press (gently) against the temples while chewing; and you can feel your masseter when you press against the back half of the lower jaw. These muscles attach to different parts of the skull and have fibres oriented different ways to pull the mandible in different directions. Our jaws can move as they do because these muscles are separated and are themselves subdivided into independently controlled parts. In addition to raising the lower jaw, for example, the back-end fibres of temporalis pull it backward, and the outer part of the masseter pulls it forward. Precise side-to-side movements are accomplished

(a)

External group

Internal group

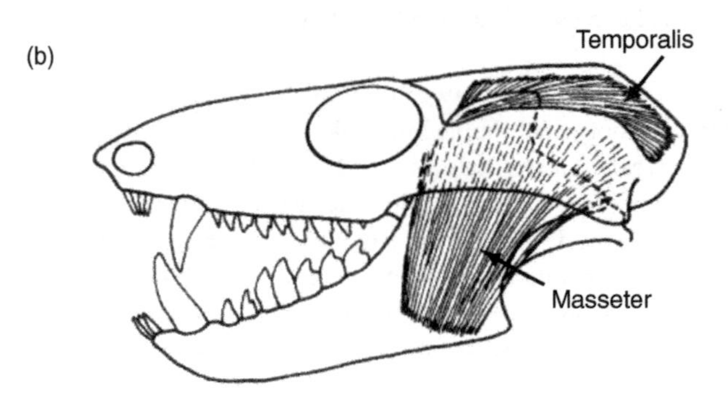

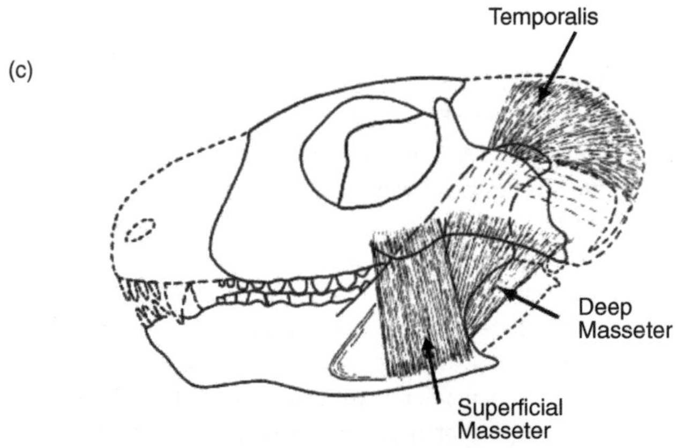

15. Synapsid skulls and chewing muscles. A, pelycosaur; B, cynodont (*Thrinaxodon*); C, advanced cynodont (*Luangwa*)

by alternating contraction of left-side and right-side chewing muscles.

Attachment sites on the fossilized skulls of early synapsids give us important clues to how these chewing muscles evolved. Recall that synapsids have a single hole, or 'window', on the side of the skull, and that this distinguishes them from other amniotes. In fact, the name synapsid comes from the Greek *syn* (together) and *apsis* (arch). The bony margins of this 'arch' provide a large surface for attachment of tendons, allowing bigger jaw-closing muscles and a more powerful, controlled bite.

Later pelycosaurs developed other important adaptations to increase the power of their bite. The coronoid eminence, a bony knob sticking up from the mandible behind the tooth row, is an important one. This knob increased the area for attachment of the external jaw-raising muscle, and moved it further away from the jaw joint and pivot point. As any seesaw rider knows, increasing distance from a pivot point means more work for a given force.

Improvements in chewing efficiency continued through therapsid evolution, with some developing large crests running front to back on top of the cranium to make more space for attachment of jaw-closing muscles, and some enlarging the coronoid eminence into a more elaborate structure, the coronoid process that mammals have today. Bony attachment sites show that the external chewing muscle divided into temporalis and masseter in the cynodonts. This, along with a reduced area for attachment of internal chewing muscles on the underside of the cranium, hints at finer control. These and other tweaks to the system over time led also to precisely balanced forces at the point of bite and reduced stress on the jaw joint. All are important for mammalian chewing.

The jaw joint. Of course, there isn't much point reorganizing the chewing muscles unless the jaw joint can handle the range of motion and forces needed for mammalian mastication. Jaw joints of most living reptiles cannot. They have a simple hinge between a bone on the bottom of the cranium, the *quadrate*, and one on the back of the mandible, the *articular*. The quadrate projects down and fits into a trough or recess in the articular. This is fine for opening and closing the mouth, but doesn't allow for much side-to-side or front-to-back motion. The mammalian jaw joint, or *temporomandibular joint* (TMJ) is very different. Not only is it made up of different bones (the *squamosal* of the cranium and *dentary* of the mandible), but the mandible fits into a recess in the cranium rather than the other way around. This allows for a greater variety of movements (see Figure 16).

Pelycosaurs and early therapsids retained the primitive hinge-like joint, but some later therapsids shrank their quadrate and increased mobility between it and the bone next to it, the squamosal. Advanced cynodonts eventually evolved a ligament connecting the squamosal to the lower jaw to help stabilize the mandible and take stress off the quadrate. Some even had contact between the squamosal and dentary, foreshadowing the mammalian jaw joint. But it still wasn't a true mammalian jaw joint. There was no projection, or condyle, on the mandible to fit into the recess, or glenoid surface, on the cranium.

The first synapsids with a true mammalian jaw joint were, by definition, the earliest mammals; the TMJ is a defining trait for the biological class, Mammalia. That's how important researchers consider the jaw joint. The very earliest of the mammals retained the old articular-quadrate connection next to the TMJ, but the new joint became increasingly dominant and ultimately replaced the old one entirely. This freed the articular and quadrate to take on new roles as middle ear bones. The articular became part of the malleus and the quadrate became the incus. Part of another lower jawbone, the angular, became the rim of the eardrum; and the evolving structure moved from the mandible to the cranium. This increased hearing sensitivity, especially at higher frequencies. But that's another story.

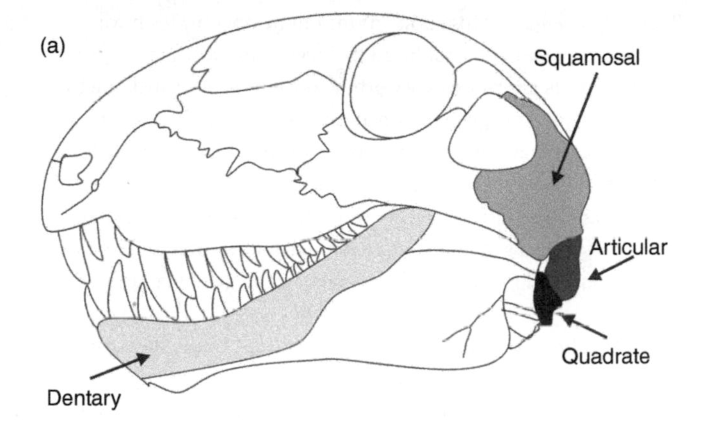

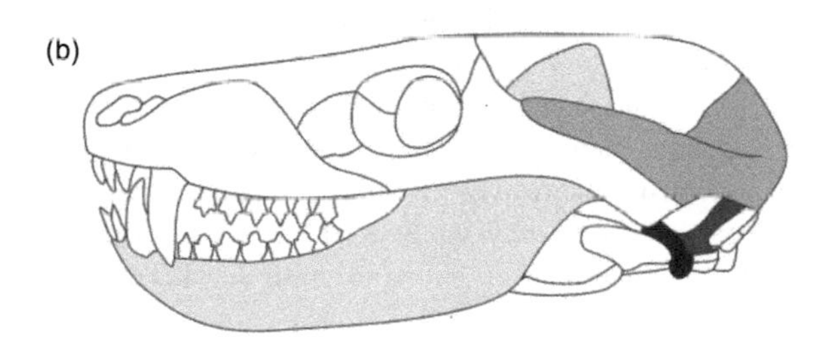

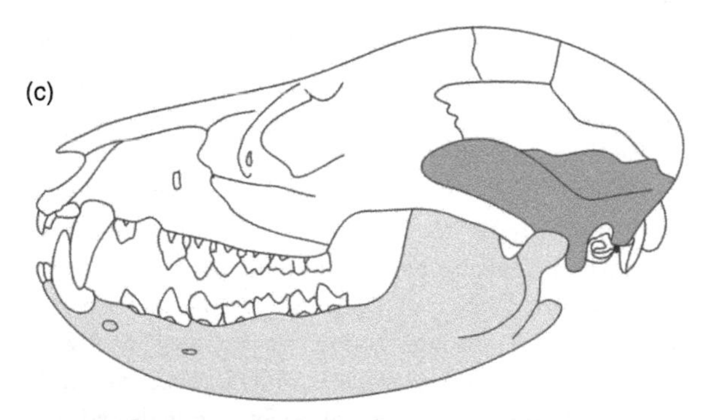

16. Synapsid skulls and teeth. A, pelycosaur (*Dimetrodon*); B, cynodont (*Thrinaxodon*); C, modern opossum

Tooth replacement. Most non-mammalian vertebrates have smaller teeth shed and replaced by larger ones as the jaw grows. Replacements alternate every other tooth or every third one to avoid big gaps. Jaw growth continues throughout life, so these animals often have many generations of teeth. Mammals work differently. They have *diphyodonty*. Incisors, canines, and premolars are usually replaced once. Molars are not replaced at all. They form as a single generation, with new ones added in succession from front to back as space in the growing jaw allows.

There are, however, some notable exceptions. Some have milk teeth that degenerate as germs and never erupt (typically all but one premolar in marsupials), erupt but are already replaced by permanent ones before birth (walruses, seals, and many rodents), or are evidently never replaced by permanent ones (toothed whales). The way they are replaced also varies. While adult teeth push the milk ones out from below in most cases, a few species, including elephants, manatees, and kangaroos, have a rather unusual replacement pattern. Their cheek teeth move forward in the jaw as if on a very slow conveyor belt, with new ones erupting from behind and old ones pushing forward until they fall out of the front of the jaw.

But why only two sets? First, mammals don't need more. Youngsters increase in size quickly, fuelled by mother's milk; and mammalian jaws stop growing in adulthood. Also, while it would be nice to be able to replace worn, broken, or diseased teeth, endless replacements and ever-growing jaws would make it difficult to keep the precise alignments between opposing teeth needed for mastication. In fact, those lizards that independently evolved precise occlusion also have fewer tooth replacements. It may come as a surprise then that diphyodonty came relatively late in synapsid evolution. Some later cynodonts probably had fewer replacements than did their predecessors, and had successive rather than alternate addition, but not even all early mammals had the modern pattern. The primitive mammal *Sinoconodon* retained multiple replacements of its front teeth, and had two sets of back ones. And its jaws continued to grow throughout life.

Hard palate. Researchers have also associated evolution of the hard palate that separates our oral and nasal cavities with evolution of the mammalian chewing system. Our long hard palate may have evolved for keeping air and food apart to prevent choking during chewing and swallowing. But there are other possible advantages to having a bony palate. It stiffens the part of the skull that holds the upper teeth, allows for greater bite forces, makes it easier to form a vacuum in the mouth for suckling and swallowing, and provides a rigid platform against which the tongue can manipulate food (think of how you mash a banana). This important structure evolved at least twice in therapsids, and became progressively more developed through cynodont evolution.

Appearance of prismatic enamel. Except for the agamid lizard, only mammals today have enamel formed in prisms. And the lizard evolved it independently anyway. Prisms offer increased tooth strength, which is important given greater stresses associated with mammalian chewing. They can also serve to create sharp edges with wear by alternating the direction in which they are laid out—hardness varies with prism orientation. Some early synapsids had column-like structures formed from enamel crystallite discontinuities; but these weren't true prisms. They lacked the interprismatic material that separates rows of prisms today. Only one cynodont has been found with true prisms, and while some early mammals had it, not all did. This raises questions as to whether it was lost in some groups, or evolved independently in different early mammals.

Mammals of the Mesozoic

Were Mesozoic mammals really small, meek insectivores cowering in the shadows of the dinosaurs? Yes, most species probably were tiny insect eaters; and they clearly didn't dominate landscapes like the dinosaurs that lived alongside them or the mammals that came later. But we are talking about 160 million years of evolution. Some early mammals were at least as big as a cocker spaniel, and they ranged from underground burrowers to terrestrial runners, arboreal climbers, semi-aquatic swimmers, and even aerial gliders. Some were herbivores, responding to the rise and spread of the angiosperms (the fruiting and flowering plants). And others were carnivores. In an interesting twist, one was even found with the remains of a dinosaur in *its* stomach!

The tale of the first two-thirds of mammalian history is actually quite complex. Mesozoic mammals evolved in what seems to be a series of successive but overlapping bush-like radiations, with later members of early groups and early members of later groups mixed together in the same deposits. Paleontologists struggle to work out how teeth evolved in these forms, especially given gaps in the fossil record and the independent appearance of some traits in unrelated species. But researchers still manage to find order in the chaos, and the fossil record gives us many examples of Nature's early experiments with mammalian tooth form. The most important of these for us is the tribosphenic molar, and we can piece together how this evolved, at least on the northern landmasses, by lining up the fossils in order from most primitive to most advanced, or derived.

The tribosphenic molar. The earliest mammals had molars not much different from those of their immediate cynodont ancestors. *Sinoconodon* and some others had three principal cusps aligned front to back with a large one sandwiched between two smaller ones. These teeth are called *triconodont*. Chewing was mostly vertical with opposing crowns sliding past one another like scissor blades, though there was a slight horizontal component to the movement. This basic arrangement served early mammals well enough so that some groups retained it, with

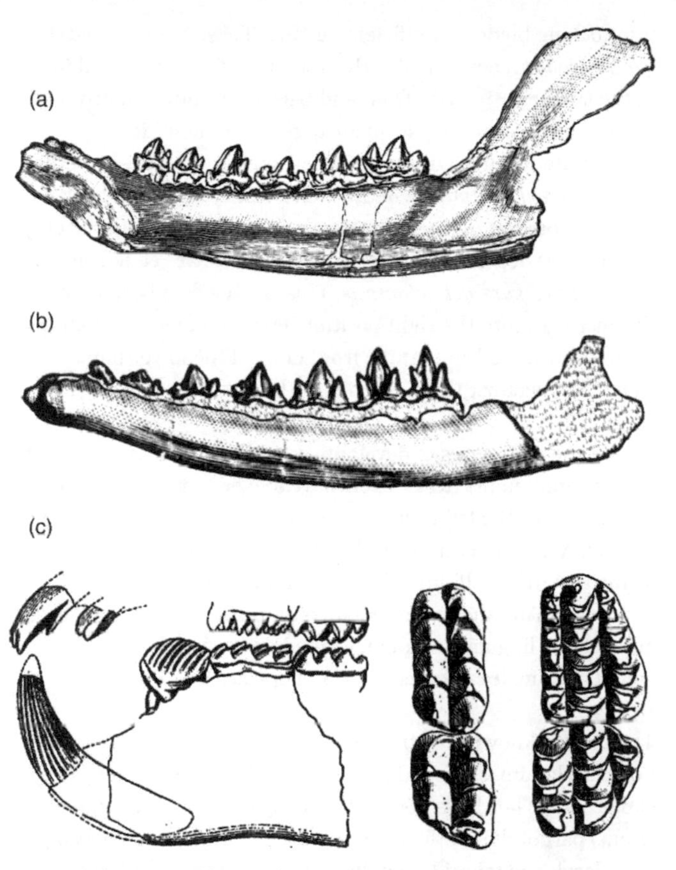

17. Mesozoic mammalian teeth. A, triconodont; B, symmetrodont; C, multituberculate

only minor embellishments, through much of the Mesozoic Era (see Figure 17).

Others had their front and back cusps rotated out of line, outward for the uppers and inward for the lowers, so that opposing rows formed reversed triangles with a zigzag pattern of interlocking blades for efficient cutting. These teeth, called *symmetrodont*, developed early and often. They are found first in the late Triassic, with front and back cusps only slightly displaced. Because they form an angle with the middle cusp greater than 90 degrees, we call species with these teeth *obtuse-angled symmetrodontans*. Over time, front and back cusps rotated more out of line, forming an angle with the middle cusp less than 90 degrees. Species with these teeth are called, aptly, *acute-angled symmetrodontans*. This is a big deal because once the cusps got into the right position, crests connecting them could be arranged so that the front ends of the lower molars sheared up against the back ends of the uppers.

We see the next series of developments in the teeth of mid to late Jurassic mammals. Researchers have looked to them to work out details of how the tribosphenic molar evolved. A new shelf, the talonid, was added on the back of the lower molars. Two cusps formed on this shelf, first the hypoconulid and then the hypoconid. And the upper molars became broader as a collar of enamel, the lingual cingulum, formed on the inside edge. This would become the protocone in later species (see Figure 2).

The earliest known true tribosphenic molars are (at least on a northern landmass) found in the aegialodontids of the early Cretaceous. They have not only shearing crests on the original (front) part of the tooth, but also a distinct protocone opposing a well-developed talonid basin for crushing and grinding. The combination of shearing and crushing parts is key to the tribosphenic molar; it makes for a great multipurpose tool to fracture foods with all sorts of material properties. And, as Cope and Osborn discovered in the late 19th century, it is from this basic type that the myriad forms of teeth in today's mammals evolved.

Truth be told, though, the fossil record is not quite so simple. We don't see a nice, neat sequence through time wherein cusps are added in front and back of the first one, those then rotate out of alignment, a platform forms behind them, and cusps are added as described. Overlapping radiations and gaps sometimes make it look like derived features appeared before primitive ones. And some fossils just leave us scratching our heads. *Ambondro*, from the mid Jurassic of Madagascar, for example, had tribosphenic-like molars much earlier than it should have, and in the wrong part of the world. The earliest known aegialodontids didn't appear until the Cretaceous and they are found in the northern hemisphere. Did the tribosphenic form evolve twice? Does this have implications for the split of lines leading to the major groups of living mammals? Time will tell as our fossil record improves.

Other Mesozoic experiments. The tribosphenic molar was an essential first step toward evolving the teeth of today's mammals. Still, it was only one of several Mesozoic Era experiments in design for crushing and grinding. This makes sense given the radiation and spread of early mammals into new adaptive zones with the new plant foods they must have offered. So, what about the other experiments? Some were very successful, but most early mammalian herbivores were probably evolutionary dead ends, too specialized to have given rise to us or any other living species.

A common theme involved cheek teeth with two or three parallel rows of cusps running front to back along the crown. Rows of cusps on opposing teeth fit between one another during occlusion. Some had mostly vertical jaw movements, with food crushed in the channels formed between the rows. Others had back-to-front jaw movements, with food ground or milled as the lowers slid along the uppers when food was interposed between them. We find this type of tooth first in the haramiyids of the Triassic. And more elaborate forms developed over time. The docodonts of the mid Jurassic, for example, developed crests to connect the rows, and enlarged platforms for crushing between opposing teeth.

The most successful and diverse group of Mesozoic mammalian grinders was undoubtedly the multituberculates, which spanned 100 million years, from the mid Jurassic to the mid Paleogene of the Cenozoic (see Table 1). These were both diverse and abundant—about half of all land mammals during their heyday. They had up to eight cusps in each of two or three rows. And there were other experiments with grinding tooth types during the late Mesozoic and early Cenozoic. The gondwanatheres of the southern hemisphere had high-crowned molars with thick enamel and rounded cusps, likely for a diet of gritty, abrasive foods. Their teeth also have distinctive crests running between the cusps of adjacent rows, and deep troughs between those crests that would have been great for milling tough items.

The age of the mammals

The Mesozoic did not end well for many species. Just over 65 mya, a meteorite about 10 kilometres across hit the Gulf Coast of Mexico's Yucatán Peninsula. Models suggest that mega tsunamis and shock waves spawned earthquakes and volcanic eruptions. Debris from the strike scattered through the atmosphere, likely bathing the world in infrared radiation, baking the surface, igniting firestorms, burning oxygen, and increasing carbon dioxide levels. Dust clouds and sulphuric acid aerosols resulting from impact on a bed of gypsum probably blocked sunlight for years, preventing photosynthesis and collapsing food chains. These things, along with massive volcanic activity from the Deccan Traps in India at about the same time and effects of dropping sea level on Earth's reflectivity and ocean currents, must have made life rather unpleasant for many creatures.

Some believe mass extinctions occurred abruptly, and others think they actually began earlier in the Cretaceous, with events at the boundary between the Cretaceous and Paleogene being the straws that broke the camel's back. Regardless, we are here and the dinosaurs are not. And all kinds of new species evolved to replace those that came before as the dust began to settle. According to paleontologist Ken Rose, about 85 new mammalian families appeared in the first epoch of the Cenozoic alone. And with new species came new adaptations and new teeth.

The Cenozoic mammalian fossil record is immense. Tens of thousands of scholarly papers have been written on the seemingly countless species. These can help us trace the origin and evolution of modern groups and how their teeth changed and diversified into the myriad distinctive forms we have today. They also offer a glimpse at dental variation in the past, and can teach us something about the evolvability of traits when we consider those that appear again and again in the fossil record. Finally, they give us a more complete picture of what Nature can do with a little embryonic tissue and some signalling proteins—novel solutions to the problems of food acquisition and processing not represented today. A few key examples illustrate these points (see Figure 18).

The evolution of today's tooth forms. The Mesozoic ancestors of the three main groups of mammals, the egg-laying monotremes (the echidnas and platypus), marsupials (kangaroos, opossums, and their kin), and placentals (other mammals), began with basic tribosphenic molars, or nearly so. Today's echidnas and adult platypus lack teeth, but the juvenile platypus has them. Primitive adult monotremes from the Cretaceous and early part of the Cenozoic of Australia and South America had molars too, with a pair of parallel crests, or lophs, running across the crown from tongue to cheek, one in front of the other. This is a common pattern among mammals, called *bilophodonty*. The fossil platypus *Obdurodon* also had bilophodont teeth, though the earliest known echidnas had already lost them. These and other fossil monotreme molars look vaguely tribosphenic—but not quite. They lack some key features, such as a functional talonid basin, and so can at best be called *pretribosphenic*.

The common ancestor of marsupials and placentals, on the other hand, likely did have true tribosphenic molars. That ancestor lived well before the Cenozoic. The earliest known placental and

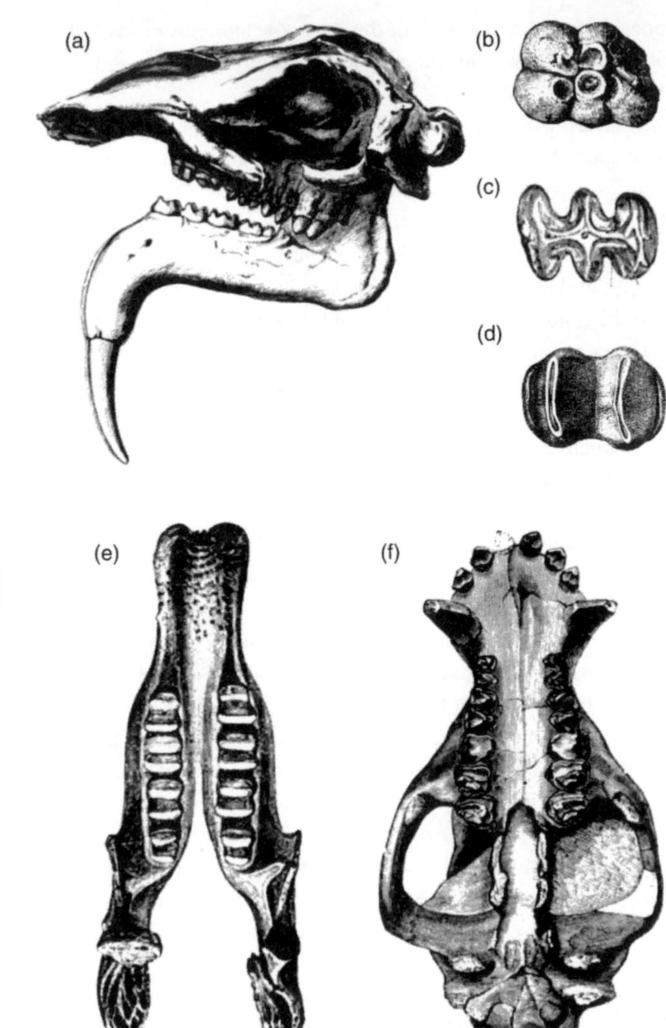

18. Cenozoic fossil mammal skulls and teeth. A, fossil elephant (*Deinotherium*); B, desmostylian; C, giant armadillo-like *Glyptodon*; D, rhinoceros wombat (*Diprotodon*); E, giant ground sloth (*Megatherium*); F, pantodont (*Coryphodon*)

marsupial come from the mid Jurassic and early Cretaceous respectively. Each has a telltale number and types of teeth; and each has subtle crown features that distinguish it from the other and relate it to later forms. Still, Mesozoic placentals and marsupials share the same basic tribosphenic pattern found in many later mammals. And we can trace several of today's marsupial and placental orders back through the fossil record to this morphological starting point. We can compare mammalian teeth, past with present, and understand how individual radiations have changed over time.

The take-home message is that there are no rules. For some, such as the opossums and shrews, teeth have remained pretty much the same through their evolutionary history. For others, they have changed dramatically. Basal rodents, for example, tended to have simple tribosphenic cheek teeth and side-to-side jaw movements, in stark contrast to today's variety of highly specialized dentitions used largely in backward-to-forward chewing. The ancestors of hoofed mammals (ungulates) had unassuming, blunt tooth crowns, whereas most today have more elaborate cheek teeth, with folded ridges running every which way, or rows of sharp, crescent-shaped crests separated by deep valleys. In fact, many fossil species did not have the distinctive features of their successors' teeth. Elephants started with simple, bilophodont molars rather than the complex ones they have today, and early rabbits lacked ever-growing cheek teeth.

Other groups went in the opposite direction. Their teeth have gotten smaller or simpler. Some have lost them completely. The earliest aardvarks had front teeth (they don't today), and the first armadillos had dental enamel (again, they don't today). Early toothed whales were heterodont (most today have peg-like teeth in front and back), and early great whales had teeth until well after baleen first appeared. We can even trace the change from tribosphenic crowns to small, peg-like teeth in the palaeanodonts, a group thought by many to be early relatives of today's toothless pangolins.

Variation in the past. Some mammalian orders seem to have more dental variation today than ever. For example, most fossil bats had tribosphenic molars with a basic W-shaped pattern of crests (see Figure 3). Few had the specialized cheek teeth of fruit and nectar eaters today. Other orders, like the primates and colugos, have about the same amount of variation today as in the past, at least for the last several million years.

Yet others varied more in the past. Fossil marsupial dental variation rivalled that today from early in the Cenozoic, especially in South America. When we add Quaternary fossil species such as the marsupial lion, with its piercing, canine-like incisors and knife-like, bladed cheek teeth, and the three-ton rhinoceros wombat, with its massive bilophodont choppers, today's variation pales by comparison. The same goes for fossil sloths and armadillos. These today have small, peg-like teeth, but not so in the past. The dog-sized horned armadillo of the mid Cenozoic had sharp, triangular cheek teeth possibly used to slice meat or tough vegetation. Two-ton Quaternary glyptodonts, also fossil relatives of the armadillo, had cheek teeth with a long crest running front to back split by three crests running side to side, all made of extra-hard dentine, perhaps for grazing. And the even larger Quaternary giant ground sloth had massive bilophodont molars with sharp-edged crests, likely for chewing leaves and other browse items. The fossil hyraxes, especially early ones, also come to mind. While there are only a handful of rabbit-sized species today, all with fairly similar teeth, hyraxes varied much more in the past. Not only did they range in size up to 1,000 kilograms, but they also had an impressive assortment of tooth shapes, from flat, bunodont forms to sharp, crescent-shaped crested ones (see Figure 19).

Common themes. We can also look to the fossil record for tooth types that evolved repeatedly in unrelated groups. Recall the common variants of the tribosphenic molar (Figure 3), those with V-shaped crests (zalambdodont), W-shaped crests (dilambdodont), and rounded, bulbous cusps (euthemorphic).

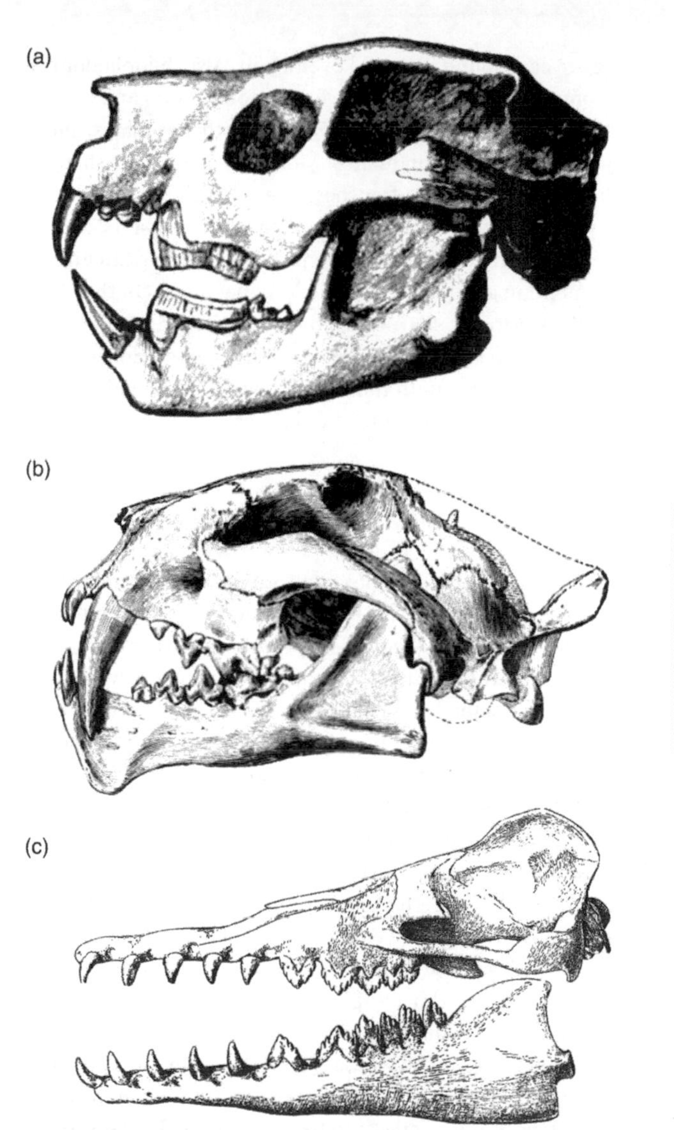

19. More Cenozoic fossil mammal skulls and teeth. A, marsupial lion (*Thylacoleo*); B, false saber-toothed cat (*Dinictis*); C, primitive whale (*Basilosaurus*)

These are each found over and over again. Also, bilophodonty is very common. We see it in some fossil monotremes and marsupials, elephants, sea cows, horses, sloths, primates, and early South American ungulates. Molars with crescent-shaped crests called *selenodont*, after the ancient Greek moon goddess Selene, are found not only in today's camels, ruminants, and hyraxes, but were common also in past South American ungulates, not to mention giant koalas and ringtail possums. On the other end of the spectrum, sharp-bladed carnassials are found in extinct South American marsupial predators (the sparassodonts), and placental carnivores, both those with no living descendants (the creodonts), and the ancestors of today's cats and dogs—albeit on different teeth.

Common themes in dental evolution are not limited to the back of the mouth. The elongated, serrated lower premolars (called *plagiaulacoid*) in brushtail possums, rat kangaroos, and bettongs today were more common in the past, and evolved in both the multituberculates and the plesiadapiforms, an order of Paleogene mammals closely related to the primates. Further forward in the mouth, saber-toothed canines appeared in several carnivorous mammals, from true and false saber-toothed cats to cat-like marsupials. For that matter, some of the larger herbivores from the early Cenozoic of North America also had sabre-like upper canines. The apatotherians, enigmatic mammals from early Cenozoic of Europe and North America, had scoop-like incisors that look like shrew teeth. And self-sharpening, chisel-shaped front teeth with thin or no enamel on the back appear again and again. Not only are they found in fossil rodents and rabbits, but in South American fossil marsupials and ungulates, the taeniodonts (specialized rooters and diggers from the early Cenozoic of North America), hyraxes, and even the giant Quaternary aye-aye (a strange-looking primate). Other groups converged on the loss of front teeth as in cows and sheep today. These include several South and North American fossil ungulates.

Evolutionary oddities. The fossil record also gives us unique teeth no longer with us today. Monotreme adult teeth are a case in point, as are the huge bladed cheek teeth of the marsupial lion. The ektopodontids, a long-lived group of fossil Australian possums, had truly bizarre molars by our standards, with two rows of up to nine cusps each running side to side, like multituberulate teeth turned ninety degrees. Then there are the lobed dentine crowns of the giant armadillos and their kin from the late Cenozoic of the Americas, which are much more complex than the peg-like teeth of their closest living relatives. There were also the desmostylians, a group of amphibious marine mammals from the mid Cenozoic that had forward-facing canine and incisor tusks, and molar crowns consisting of columns, or pillars of enamel, bound together like cylindrical honeycombs. No one has teeth like that today.

Chapter 6
Mammalian teeth today

Mammalia is an amazingly successful and diverse biological class. From the bumblebee bat, lighter than the smallest coin in your pocket, to the behemoth blue whale, as heavy as a Boeing 747 airliner, mammals burrow, swim, crawl, hop, run, climb, glide, and fly through a fantastic variety of habitats. They range from Arctic tundra to Antarctic pack ice, the ocean's depths to high mountain peaks, and open desert to dense rainforest. Some are herbivores, and eat grass or browse on parts of other plants. Some eat fungi. Others are carnivores, with prey ranging from microscopic plankton to the largest animals on the planet. Some are picky, and concentrate on just a few foods, and others will eat almost anything they can get their mouths around.

What is the key to this remarkable diversity? If you're thinking 'teeth', you're on the right track. The key is actually endothermy, our ability to heat the body from within. This is more than just being warm-blooded; it is creating heat from food. As Tom Kemp has written, 'Nothing is more fundamental to the life of mammals than their endothermic temperature physiology.' Mammals can live in colder climates and places with more fluctuating temperatures, and be active during the cool, dark night. Endothermy means more controlled conditions for chemical reactions in the body, so more complex systems can develop. And it permits sustained activity and higher travel speeds for larger territories and greater migration distances, stamina for foraging, predator avoidance, and parental care. It also allows for longer periods of growth and development of energy-hungry tissues, such as the brain. Without endothermy, mammals couldn't be mammals.

But endothermy isn't cheap. It takes a lot of energy to run the body's furnace; and the more extreme the air (or water) temperature, the more it takes. A mammal at rest typically guzzles fuel at a rate five to ten times that of similar-sized animals that rely on their surroundings for heat (ectotherms); and rates can climb to ten to fifteen times higher with heavy activity. Mammals must wring as many calories as possible from the foods they eat. And this is where teeth come in. Teeth rupture protective casings such as insect exoskeletons and plant cell walls to release nutrients that would otherwise pass through the gut undigested. They also fragment items to increase exposed surface area for digestive enzymes to act on; more surface area usually means access to more energy. Could this be accomplished without teeth? Certainly. Birds grind food with gastroliths, small stones in their muscular gizzards. Even some toothed animals, such as crocodiles and mammals including seals, sea lions, and porcupines, are occasionally found with stones in their stomachs. But teeth are clearly the mammalian solution to food grinding. And while a few mammals manage without them, teeth are an inseparable part of the mammalian identity, both for individual species and for the biological class as a whole.

Nature puts intense pressure on teeth, literally and figuratively, to provide mammals with access to the energy they need for endothermy. But internal heating also means more food options because mammals can live, and look, in more places. And because carbohydrate, protein, and fat can all be tapped for fuel, mammals can find something to eat just about anywhere. We can understand mammalian diversity and the diversity of mammalian teeth in terms of demand for energy, variety of items on the biospheric buffet, and improved access that comes with highly evolvable teeth.

Before we can begin exploring dental diversity in today's mammals, though, we need a way of organizing the more than 5,000 species (see Table 2). I use the classification described in Don Wilson and DeeAnn Reeder's *Mammal Species of the World* combined with recent studies of relatedness based on genetic similarity. There are three principal groups: Protheria, Marsupialia, and Placentalia. The protherians are the egg-laying monotremes, again, the platypus and echidnas. We can sidestep these entirely here because none today has teeth, at least not as an adult. The marsupials and placentals, on the other hand, give us plenty to discuss.

Table 2. Classification of the living mammals

Subclass/ Infraclass	Supraorder	Order	Common name
Protheria		Monotremata	platypus, echidnas
Marsupialia		Didelphimorphia	opossums
		Paucituberculata	shrew opossums
		Microbiotheria	monito del monte
		Notoryctemorphia	marsupial moles
		Peramelemorphia	bandicoots, bilbies
		Dasyuromorphia	Tasmanian devil, quolls, dunnarts, numbat
		Diprotodontia	kangaroos, wallabies, possums, koala, wombats

Subclass/ Infraclass	Supraorder	Order	Common name
Placentalia	Xenarthra	Cingulata	armadillos
		Pilosa	sloths, anteaters
	Afrotheria	Afrosoricida	golden moles, tenrecs
		Macroscelidea	elephant shrews
		Tubulidentata	aardvarks
		Hyracoidea	hyraxes
		Proboscidea	elephants
		Sirenia	dugongs, manatees
	Laurasiatheria	Cetartiodactyla	whales, even-toed ungulates (ruminants, camels, hippos, pigs peccaries)
		Perissodactyla	odd-toed ungulates (horses, tapirs, rhinos)
		Chiroptera	Bats
		Carnivora	cats, civets, hyenas, mongooses and kin, bears, dogs, weasels and kin, seals, walrus
		Pholidota	pangolins
		Eulipotyphla	hedgehogs, gymnures, shrews, moles, solenodons
Euarchontoglires		Scandentia	tree shrews
		Dermoptera	colugos

(*continued*)

Table 2. Continued

Subclass/ Infraclass	Supraorder	Order	Common name
		Primates	primates
		Lagomorpha	hares, rabbits, picas
		Rodentia	rodents

Details can be found in Wilson and Reeder's *Mammal Species of the World: A Taxonomic and Geographic Reference* (3rd edition) and Ungar's *Mammal Teeth: Origin, Evolution, and Diversity*

Marsupials

The marsupials are more than kangaroos and koalas. There are hundreds of species, grouped into seven orders—three in the Americas and four in Australia and islands of the South Pacific. Marsupials dig, walk, hop, and glide their way through a remarkable range of habitats. These include carnivores, insectivores, fungus eaters, and a variety of herbivores, from generalists to those that specialize on grasses, fruits, leaves, roots and tubers, or nectar and pollen. And they have dental variation to match. As paleontologist Mike Archer has said, it is enough to give one 'pouch envy'! Marsupialia is a fascinating, diverse group that offers great examples of what Nature can make from simple, peg-like front teeth and tribosphenic back ones.

Marsupials have by tradition been divided into polyprotodonts, with four or five small, peg-like incisors and a canine in each of the four quadrants of the mouth, and diprotodonts, with fewer front teeth, often a pair of large, projecting ones in the upper and especially the lower jaw (see Figure 20). Most American and Australasian faunivores, such as the opossum and Tasmanian devil, have been classified as polyprotodonts, whereas the American shrew opossums and Australasian herbivores, such as the kangaroos and koalas, have been called

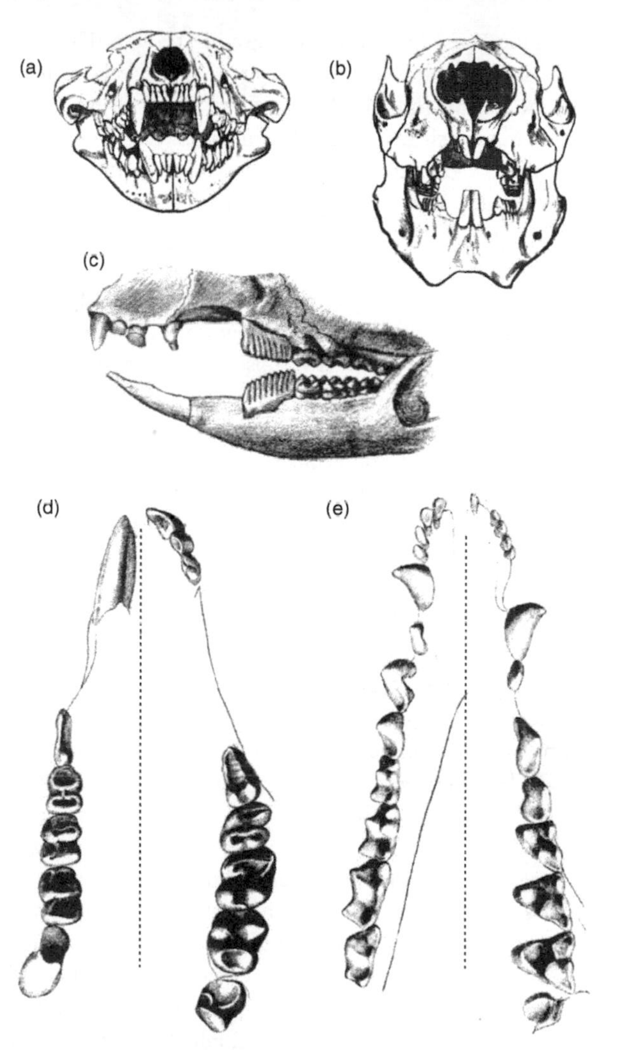

20. Marsupial teeth. A, polyprotodont (Tasmanian devil) and B, diprotodont (koala) types. C, a musky rat kangaroo in side view (note the plagiaulacoid premolar). Lower (left) and upper (right) teeth of D, a kangaroo and E, an opossum

diprotodonts. These are not natural groups, though. The Australasian herbivores are actually more closely related to many of the polyprotodonts than they are to the shrew opossums. In fact, shrew opossum and kangaroo lower incisors aren't even evolved from the same ancestral tooth types; they have converged from the primitive second and third incisor respectively.

Besides, there is much more to marsupial dental diversity than big and little incisors. Their cheek teeth show an extraordinary variety of forms that in many ways mirror those of placental mammals. This is especially impressive when you consider that there are fifteen times as many placental species as marsupials. It also shows that Nature has certain consistent answers to the question 'What happens when the primitive mammalian molar is given time and ecological opportunity?' Marsupial moles are like placental golden moles, with sharp, zalambdodont molars bearing Λ-shaped crests set off on a high shelf. These mammals share a penchant for insects and other invertebrates living underground. The more omnivorous opossums have dilambdodont molars, with double Λ-shaped crests (resembling a W) on that shelf, like those of placental desmans, shrews, trees shrews, and most bats. Carnivorous Tasmanian devils and quolls have sharp, notched carnassial-like cheek teeth resembling those of dogs and cats, albeit in different tooth positions.

At the opposite end of the spectrum, leaf-eating koalas and ringtail possums are selenodont, with crescent-shaped crests on their molars, like those of camels and cows. And wombats have ever-growing, figure-8-shaped rims of enamel for grinding grit-laden vegetation, much like the premolars of burrowing pocket gophers. Kangaroo molars are bilophodont, with two rows of cusps, not unlike those of tapirs and some primates and rodents. Speaking of kangaroos, several species share with elephants and manatees the conveyor-belt-like horizontal replacement of cheek teeth from back to front. This is especially impressive in the pygmy rock wallaby, which can erupt up to about nine extra molars in each quadrant over a lifetime.

Marsupials sometimes have unique dental adaptations too. The termite-eating numbat has up to fifty-two small teeth in the mouth at once, including buccolingually compressed premolars that look much like the primitive tricondont cheek teeth of early Mesozoic mammals. Some possums and rat-kangaroos have serrated, blade-like premolars used for processing tough or hard foods such as straw, nuts, and beetles. These teeth look just like those of fossil multituberculates and plesiadapiforms. The honey possums, on the other hand, have reduced their teeth to tiny pegs, though they do have other interesting feeding adaptations, such as a bristled tongue for lapping nectar and pollen from flowers.

Placentals

The placentals are even more impressive. These are spread over eighteen orders grouped using genetics-based features into four supraorders, two that appeared first in the southern hemisphere (Xenarthra and Afrotheria) and two from northern continents (Laurasiatheria and Euarchontoglires). These vary greatly in number of species and diversity of dental adaptations. Each teaches us something different about relationships between species richness, diet, and teeth.

Xenarthra. Xenarthrans are the sloths, armadillos, and anteaters. They make up less than 1 per cent of placental mammal species. Most live in South and Central America, though we have our share of armadillos up here in the Ozark Mountains of Northwest Arkansas. Their range of habitats, from underground to the trees, is modest compared with other placental supraorders, and their dental variation is unimpressive. But the variety of different diets they have, given a lack of ornate teeth, is remarkable. They teach

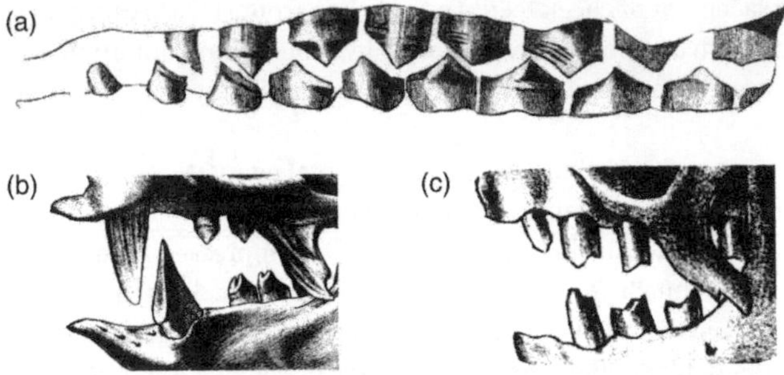

21. Xenarthran teeth. A, armadillo; B, two-toed sloth; C, three-toed sloth

us that some mammals can make their way in the world just fine without elaborate dental toolkits (see Figure 21).

Anteaters eat insects and have no teeth, but they do have long, thin snouts and tongues, and imposing claws for breaking into ant and termite nests. Armadillos also have a penchant for insects, though individual species consume varying amounts of other animals and plant parts. And sloths eat mostly leaves. Armadillos and sloths tend to lack distinct front teeth, and their cheek teeth are simple, single-rooted pegs—though their crowns can wear to form bevelled or chisel-shaped surfaces. These teeth are ever-growing and in adults lack enamel, but they often have a highly mineralized, hardened outer layer of dentine, sometimes covered in cementum.

Afrotheria. The afrotherians make up less than 2 per cent of placental mammalian species, but show big ecological diversity in a small package and have an assortment of teeth to match. Afrotherians range from tiny, shrew-like tenrecs, weighing only five grams, to the largest elephant, at more than ten tonnes. Most species live in Africa, but hyraxes and elephants are also found in Asia, and sirenians (manatees and dugongs) live in tropical waters of the Atlantic and Indopacific as well as river systems in the

Americas and Africa. Afrotherian habitats range from subterranean to terrestrial to arboreal, and from freshwater to marine. Their diets are also diverse. Golden moles, tenrecs, aardvarks, and elephant shrews are all principally insectivorous. Hyraxes, elephants, dugongs, and manatees, on the other hand, are herbivores. Some prefer grass, others favour tree, bush, or forb parts. Yet others are mixed feeders with flexible diets.

So what about their teeth? Some, such as the aardvarks and manatees, have no front teeth, but others, such as the golden moles and some tenrecs and elephant shrews, have the full primitive placental complement of three incisors and a canine in each quadrant (see Figure 22). Front teeth range from simple, peg-like structures in elephant shrews to chisel- or shovel-shaped incisors in hyraxes and tusks in dugongs. The tusk of the elephant, especially the African bull, is most impressive. This modified upper second incisor can reach nearly 3.5 metres in length. It erupts with a thin layer of enamel that wears away quickly, leaving only dentine, or ivory, on the surface. You can tell elephant ivory by its unique cross-section—intersecting lines spiral out from the pulp chamber to form a checkerboard pattern. Manatees, on the other hand, have swapped adult front teeth for dental pads made of keratin, like your fingernails. These pads are great for cropping sea grasses and grasping other vegetation.

Afrotherian cheek teeth also run the gamut. The aardvark and dugong have simple pegs, and the golden moles and some tenrecs are zalambdodont. Other tenrecs are dilambdodont. Elephant shrews and hyraxes have square-shaped, or *quadrate*, cheek teeth, and the latter often have crescent-shaped, selenodont crests running front to back across each cusp. Manatee cheek teeth tend to be bilophodont, with two rows of cusps, often three in each row. And elephants have complex crowns called *loxodont*, with about five to twenty-nine parallel ridges or plates running buccolingually across the crown. Recall that elephant and manatee cheek teeth have, like kangaroos, horizontal tooth

(a)

(b)

(c)

(d)

(e)

22. Afrotherian teeth. A, tenrec (uppers); B, hyrax lowers (left) and uppers (right); C, manatee (lowers); D, aardvark (lowers); E, elephant (lower tooth)

replacement. These move forward about 1 millimetre per month in manatees, with three dozen passing through each quadrant in a typical lifetime.

Laurasiatheria. Laurasiatheria offers the ultimate example of what Nature can accomplish starting with a small, primitive insectivore. The order comprises more than 2,200 recent species, from whales and pigs to camels and cows, horses and rhinos, dogs and cats, moles and shrews, bats, and pangolins. These include some of the most conservative and some of the most specialized mammals on the planet, from the smallest of aerial bats weighing 1.7 grams, to the largest of aquatic whales at 170,000,000 grams. Laurasiatherians are spread from the northern Arctic sea to Antarctic pack ice and most places in between. They live in seemingly countless habitats. You can find them in the air and trees, on and under the ground, and in freshwater and marine ecosystems. And their diets are unrivalled in variety among the mammals. Some are extreme specialists and others are generalists. Some are strict herbivores, and eat grass, browse, or both. Some consume fungi or nectar. Others are faunivores, and prey on animals of nearly every size and shape, from zooplankton to blue whale.

Along with this variety of diets comes a remarkable radiation of tooth forms and other feeding adaptations (see Figures 23 and 24). For example, shrews have thin, curved, forceps-like incisors used for catching and holding small prey. The solenodon, a burrowing shrew-like mammal from Cuba and Hispaniola, has a big, sharp lower second incisor with its enamel folded over to form a partially enclosed tube through which venom is injected into small prey. And vicuñas, llama-like creatures in the high Andes, have ever-growing and chisel-like incisors, not unlike those of rodents and rabbits, used to crop small forbs and grasses close to the ground. None of these can compare, however, to the incisor tusk of the narwhal of the north Atlantic and Arctic Oceans. Tusks are most common in male narwhals, though not all have them.

(a)

(b)

(c)

(d)

(e)

23. Laurasiatherian teeth. A, camel (uppers); B, rhinoceros (uppers); C, orca whale; D, horse lowers (left) and uppers (right); E, pig lowers (left) and uppers (right)

And they usually have only one, a left upper incisor that can grow up to 3 metres long. The narwal tusk lacks enamel, but it is covered in cementum, and has a spiralling groove, like a unicorn's horn. It also has millions of nerve endings, used to detect changes in water temperature, pressure, and chemistry.

Canine tusks are also common in Laurasiatherians. Several deer-like species and pigs have them. They are commonly larger in males, and are often used for display and fighting. Hippopotamus and walrus canines are sometimes more than a metre in length. The Indonesian pig-like babirusa has long, curved upper and lower canine tusks; the uppers also grow upward in a backward arch—in older individuals they curve around so far that they contact the forehead.

There are other examples too. The strap-toothed whale has a long, thin pair of ribbon-shaped tusks rising from the mandible and wrapping up around the head to nearly close off the mouth, though not enough to keep their favourite food, squid, out. As is common for tusks, these are used for display and fighting.

Camels and ruminants are at the opposite end of the spectrum, with a keratin plate replacing the upper front teeth. The lower incisors bite against this dental pad to bring just the right pressure to 'comb out' soft, weak grass leaf blades and other nutritious plant parts but leave behind stronger, low-quality stems.

Laurasiatherian cheek teeth are also amazingly variable. Some species are conservative, and retain the basic tribosphenic form, such as the zalambdodont solenodons, and the dilambdodont bats, shrews, and moles. Others, such as hedgehogs, have quadrate molars, and tapirs are bilophodont. Pigs and hippos typically have blunt teeth with four principal cusps, but crown surfaces are often wrinkled and complex with up to thirty tiny cusps called *cuspules*. Some have highly modified cheek teeth, like the bladed carnassials of dogs and cats. The upper last premolar and lower first molar each has a sharp blade running anteroposteriorly and meeting in the middle of the tooth. The lower V-shaped blade slides up against the upper Λ-shaped one to keep food from spreading as it's sliced. The camels, deer, giraffe, and cow all have selenodont molars. Each cusp, two in the front and two behind for most, has a crescent-shaped crest running front to back. These wear quickly, forming parallel rows of sharp edges where enamel meets dentine for shearing tough vegetation. And the rhino and horse have tightly packed and elaborately folded rims of enamel, making sharp crests with wear that twist about the crown. In the horse, the combined length of those crests is four times the circumference of the tooth itself. These make excellent surfaces for grinding tough vegetation.

On the other hand, some laurasiatherians have simplified teeth, such as the cone-shaped structures of some sea lions and seals and the peg-like ones of many toothed whales. Dolphins can have up to 260 of these in the mouth at once. Other seals have unusually developed cusps, with cheek teeth resembling dagger-like tridents or bizarrely hooked structures forming sieves to filter krill (see Figure 24). Pangolins, in contrast, have lost their teeth entirely but, like anteaters, have developed long snouts and sticky tongues for grasping ants and termites. The great whales have also lost their teeth, but have rows of triangular baleen plates that hang like parallel combs from each side of the palate. These have brush-like bristles with overlapping fringes that form giant mats, not unlike the air filter in your car or home, for trapping small fish, krill, and plankton. These keratin structures have nothing to do with teeth, but as Darwin wrote, are among the whale's 'greatest peculiarities'.

Laurasiatherians also offer great examples of diet-related differences in teeth of closely related species. Frugivorous bats have blunter cheek teeth than insectivores, and nectar feeders have tiny, simple teeth. The vampire bats are very specialized, with huge, sharp canines and upper incisors for piercing their prey, but greatly reduced cheek teeth. And among the carnivorans, compare

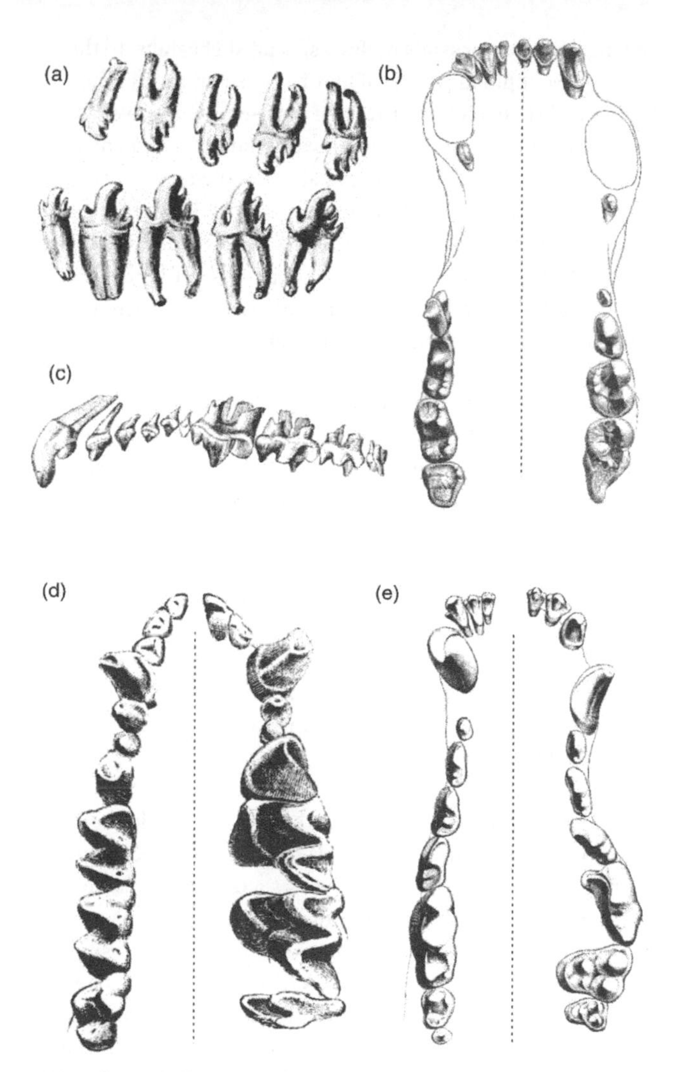

24. More laurasiatherian teeth. A, crab-eater seal; B, bear lowers (left) and uppers (right); C, shrew (uppers); D, bat lowers (left) and uppers (right); E, wolf lowers (left) and uppers (right)

the long, sharp carnassials of flesh-specialist cheetahs, to the blunter, more bulbous cusps of herbivorous pandas and the strong, thickly enamelled crowns of bone-crunching hyenas. More subtle differences are found within families. Grazing antelopes that chew a lot and have abrasive diets tend to sport higher-crowned cheek teeth than do browsers. Proportions of shearing and crushing areas on molar surfaces also vary between fossas, cat-like animals from Madagascar, and between species of New World leaf-nosed bats, mongooses, bears, and weasels—all related to subtle differences in their diets.

Euarchontoglires. The final supraorder, Euarchontoglires, is the most species-rich. It includes about 60 per cent of all mammalian species. These are the rodents, rabbits, tree shrews, colugos, and primates. They are widely distributed, especially the rodents, and can be found in a tremendous variety of habitats from the Arctic to the Subantarctic, underground to tree canopy and aquatic environment to desert. That said, they vary only modestly in diet compared with laurasiatherians. Most are small herbivores, but some prey on insects and other invertebrates, or small vertebrates. Some are dietary specialists, but others are adaptable opportunists, and eat a variety of plant parts and small animals as seasons and local availability allow.

Euarchontoglirans also vary less in their teeth than do the other northern continent placentals (see Figure 25). Considered another way, though, think about how successful they have been with subtle variants on just a few dental themes. There are more than 2,000 species of rodent alone, yet most have simple chisel-shaped incisors and flat cheek teeth with folded enamel rims. That said, euarchontogiran teeth do vary somewhat, and in some interesting ways. Tree shrews have small and simple pointed front teeth, whereas rodents and rabbits have large, ever-growing, and self-sharpening incisors, like wombats and vicuñas. In fact, self-sharpening incisors are called *gliriform* after rodents and rabbits, which are grouped together in the grand order Glires.

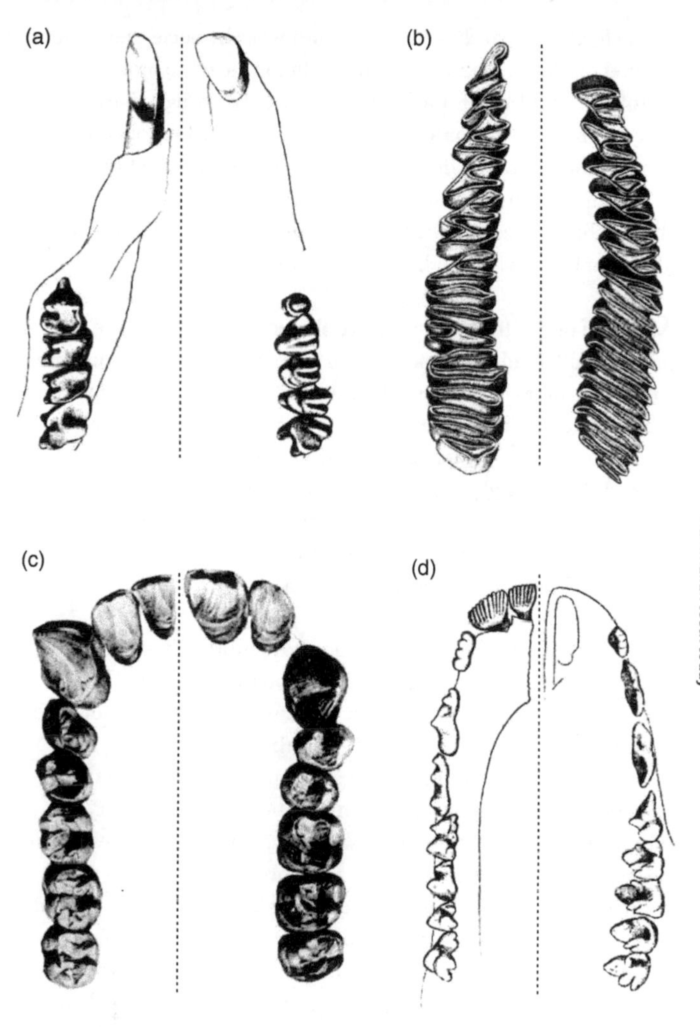

25. Euarchontogliran teeth. A, marmot; B, capybara; C, chimpanzee; D, colugo. Lowers are on the left and uppers are on the right for each pair

These form a sharp chiselled edge with wear because they lack enamel on their inner surface. But colugos have the most distinctive front teeth. Each has up to twenty prongs, like a tiny hair comb. Some primates have long, thin lower incisors and canines that together also form a functional comb used in grooming their neighbours, whereas others have broad, shovel-shaped incisors for husking fruits. And the aye-aye has gliriform incisors, much like its fossil forebear.

As for the rest of the tooth row, tree shrews and colugos have primitive dilambdodont molars, and primates often have quadrate, sometimes bilophodont, ones typically with three to five cusps. Many rabbits and rodents have crowns with simple dentine surfaces rimmed by enamel, whereas others have elaborately folded bands of enamel. These bands often push inward from the cheek and tongue sides and, in some cases, meet in the middle to form transverse plates of enamel like those of elephant teeth. And crowns vary from short to tall to ever-growing (hypselodont).

Euarchontogliran cheek teeth often show subtle differences related to diet. Rodents that consume tough vegetation tend to have more complex occlusal surfaces than those that do not. The capybara is the most interesting—its third molars are exceptionally long and have nine or ten parallel plates on the uppers and six on the lowers. Primate cheek teeth also vary by diet. Leaf eaters and insectivores have longer shearing crests than do frugivores. Still, you would be hard pressed to argue that Eurachontoglires has much dental diversity, especially in light of its very broad distribution and species richness. One would think that the combination of high reproductive rates, short generation lengths, geographic spread, and habitat variation, especially in rodents, should be a recipe for an extraordinary radiation of tooth forms. But we just don't see it. Perhaps the pairing of ever-growing, self-sharpening incisors and flat cheek teeth with bands of enamel and dentine simply makes for a good way of processing a very broad variety of food types.

Recurrent themes and unique solutions

To me the most extraordinary thing about mammalian teeth is the repeated occurrence of some forms in unrelated species. Nature comes up with the same solutions to the fundamental challenges of food acquisition and processing again and again; and mammalian teeth provide some of the finest examples of convergent evolution. These examples offer us insights both into relationships between form and function, and into how the genes that code for tooth shape change and express themselves. Again, gliriform incisors are found in groups as different as wombats, vicuñas, rodents, and aye-ayes. Perhaps it isn't surprising that some of the more primitive tribosphenic molar forms, such as zalambdodont and dilambdont types, show up repeatedly. But what about selenodonty in koalas and cows, loxodonty in elephants and capybaras, and figure-8-shaped, ever-growing cheek tooth crowns in wombats and pocket gophers?

At the opposite end of the spectrum, some species have reverted to simple, cone-shaped teeth, or have even lost their teeth entirely through the course of evolution. Honey possums, aardvarks, armadillos and sloths, walruses, dugongs, and dolphins all have simple, homodont pegs. Some have enamel caps, and others do not. Yet others start with thin layers of enamel that wear away soon after the teeth erupt. And echidnas, anteaters, and pangolins have all lost their teeth through evolutionary history, but share long, narrow snouts and long, sticky tongues for catching colonial ants and termites.

There are other tooth forms that are more unique, but just as interesting. Recall the venom syringe of the solenodon, the unicorn-horn-like sensory tusk of the narwhal, and the comb-shaped front teeth of the colugo. The vampire bats have massive, piercing canines and upper incisors, and their cheek teeth are small, but have wedge-shaped cutting edges, the back ones forming serrated blades. The crab-eater seal has long, hooked cheek tooth lobes that loop around and interdigitate in occlusion to form a sieve for straining krill. Hippo molars also have distinctive cusps, each of which has three lobes, which wear to a rim of enamel, taking the shape of a three-leaf clover. And we mustn't forget the elaborate, twisting bands of enamel covering the crowns of horse cheek teeth, or the elongated third molar of the warthog, which can have dozens of tiny cusps.

Chapter 7
Human teeth and their history

Open your mouth and look in a mirror. Not very impressive, are they? Your teeth are small, flat, and boring compared with the ornate dentitions of cows, horses, dogs, and cats. To make matters worse, your teeth have probably had cavities, and some may have erupted crooked or not at all. Millions of us suffer fillings, crowns, wisdom tooth extractions, and braces each year. Most other species don't have such widespread dental disease and orthodontic disorders. Why are we so different? The answer is rooted in our evolutionary history.

The hominin fossil record

Studies of genetics teach us that the lines leading to humans and our nearest living relatives, the chimpanzees, split during the Neogene, at least 7–8 mya. All species on our side of that split, whether our ancestors or evolutionary side branches, are called *hominins.*

While paleoanthropologists don't all agree on how to divide up hominin fossils into species, or how they are related to one another, they can be arranged reasonably into four basic groups (see Figure 26). The first group includes what many believe to be the earliest hominins, dated from 6 or 7 to 4.4 mya. These are *Sahelanthropus* from Chad, *Orrorin* from Kenya, and

Ardipithecus from Ethiopia. We can call them ardipiths for short. The second group includes species of the genus *Australopithecus*, between about 4.2 and 2 mya. These hominins have been found in Chad, eastern Africa, and South Africa. *Australopithecus* species are also called gracile australopiths because their skulls and jaws are relatively slender compared with those of the third group, *Paranthropus*. *Paranthropus* species, the robust australopiths, are known from eastern Africa and South Africa and date to between about 2.7 and 1.2 mya. The earliest known fossils in the fourth group, our genus, *Homo*, appear around 2.4 mya. Early *Homo* also lived in both eastern and southern Africa and, then, around 1.8 mya, expanded into Asia and much later, the rest of the world. We can envision the human family tree like the letter Y, with the first and second groups on the main line leading to humans, and the third and fourth groups marking an evolutionary fork in the road around the time savannas spread across eastern and southern Africa. The robust australopiths are an offshoot, a group of specialist hominins that lived alongside *Homo* for the first half of the evolutionary history of our genus.

What did the teeth of the earliest hominins look like? And how did they change over time to become different from those of other apes? These are common questions for paleoanthropologists, especially since we have more teeth to work with than any other part of the body.

Tooth size. The first thing most people notice when comparing our teeth with those of chimpanzees is the canines. Other apes have long, sharp upper canines with a back edge that hones, or sharpens, against the front end of the opposing premolar. This is especially true of males, which have bigger canines than females. We call this *sexual dimorphism*. Charles Darwin reasoned that the difference between the sexes in canine size evolved for threat display and combat between males for access to mates. Humans have less of a difference between males and females, and these teeth are shorter and do not hone. This

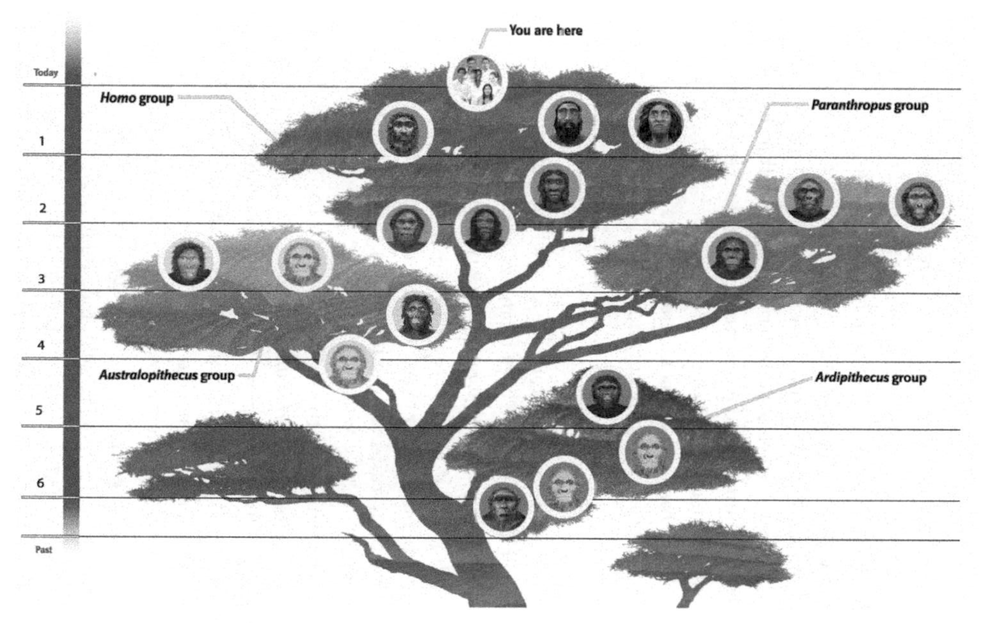

26. The human family tree. You can identify individual species on the Smithsonian Institution's website, http://humanorigins.si.edu/evidence/human-family-tree

may reflect changes in social structure during the course of human evolution—and the fact that we tend not to bite one another when competing for mates.

Ardipith canines are about the same size as of those of female chimpanzees; small compared with most living apes. They also differ less between males and females, and did not hone. Their canines are still larger than ours, but they make clear that changes in size, contact with opposing teeth, and differences between the sexes were well under way shortly after the divergence of human and chimpanzee lineages. Hominin canines were smaller in *Australopithecus*, and by the time we get to *Paranthropus* and *Homo*, these teeth are essentially like ours, hardly projecting beyond the incisors or other teeth in the row.

Fossil hominins also differ from one another and living apes in the sizes of their cheek teeth. Cheek tooth size has been related to food energy yield. The idea is that larger chewing platforms are needed if low-quality food is eaten because the body requires more of it. This may explain why, for example, leaf-eating gorillas have larger cheek teeth than fruit-eating chimpanzees.

Ardipith postcanine teeth tend to be slightly larger than those of chimpanzees, and *Australopithecus* ones are larger still. *Paranthropus* has the largest of all hominins, with molars having up to five times the occlusal surface area of ours. Earliest *Homo* also had fairly large molars, but they have become smaller, species by species, since. So, for the first half of hominin evolution our ancestors' molars got larger, but then they got smaller. At first glance, this suggests a decrease in food quality (or at least the need for a large chewing platform) during the first half of hominin evolution, followed by an increase in quality through the evolution of *Homo*. Perhaps tools used to prepare food and ultimately cooking also lessened the need for large chewing platforms.

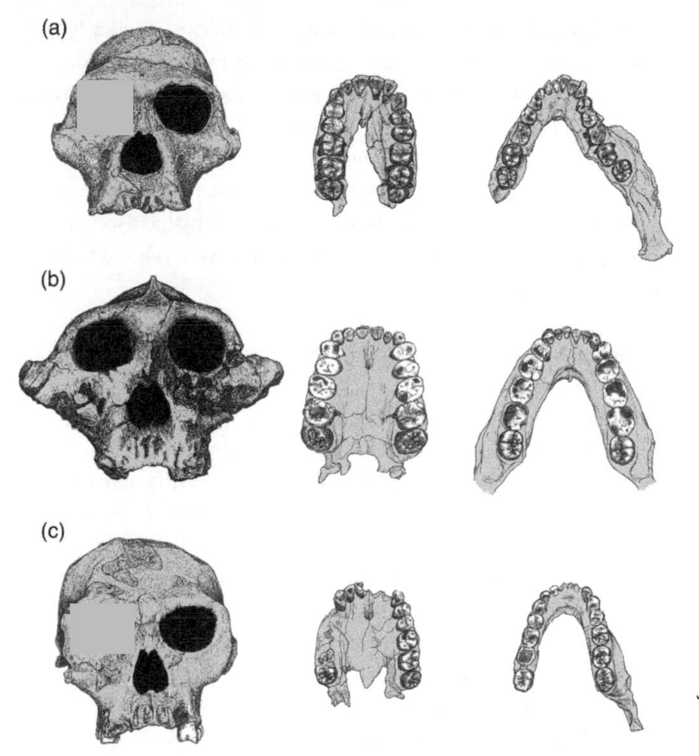

27. Early hominin skulls and teeth. A, *Australopithecus*; B, *Paranthropus*; C, early *Homo*

But why would our ancestors' teeth shrink? Isn't bigger better? Laboratory studies by biomechanics researcher Dan Lieberman and his colleagues have shown that animal jaws grow longer with heavy use. We also know that tooth size and jaw length are matched. If the jaw is too small, malocclusion and impacted wisdom teeth can follow, both of which can cause serious problems. So, as biological anthropologists James Calcagno and Kathleen Gibson have suggested, less chewing may have led to wimpier jaws and smaller teeth to match (see Figure 27).

Tooth shape. Sharper teeth with longer crests can efficiently shear or slice tough foods, such as leaves or meat, whereas blunter ones without fragile crests can withstand breaking given high forces used to crush hard objects, such as nuts or roots. This explains why gorillas, which sometimes rely on tough leaves and stems, have sharp, pointy molar teeth with longer crests than do chimpanzees. It also explains why hard-object-feeding monkeys have flatter molars than closely related fruit eaters or leaf eaters.

All the hominins had flat teeth, at least compared with a gorilla. The crowns of australopiths are flatter than those of chimpanzees at comparable stages of wear, and *Paranthropus* cheek teeth are flatter than those of *Australopithecus*. Gracile and especially robust australopith teeth were clearly able to resist breaking with heavy use. Interestingly, early *Homo* teeth are somewhat sharper than those of their australopith predecessors and contemporaries. Perhaps this allowed them to more efficiently shear tough foods, such as meat. In fact, animal bones with the telltale cut marks of butchery are found at sites with early *Homo* fossils. Undisputed evidence dates back to around 2.5 mya, and cut-marked bones appear in large concentrations after about 2 mya.

Tooth structure. Chimpanzee teeth are organized differently than ours. They have a thin coat of enamel, whereas ours is thicker, at least relative to the amount of underlying dentine. Some have argued that thick enamel evolved to extend the life of a tooth worn by abrasive or grit-laden foods when our ancestors came down out of the trees, especially if those foods required a lot of chewing. On the other hand, the most terrestrial of the living great apes, the gorillas and chimpanzees, actually have thinner enamel than the more arboreal orangutans, as Richard Kay has pointed out. Perhaps, then, thickened enamel evolved to strengthen teeth so they would not break while crushing hard foods, such as nuts or roots.

While ardipith enamel thickness varies within and between teeth and species, they tend to have thicker enamel caps than those of chimpanzees and gorillas but thinner ones than those of later hominins. Had they begun to evolve thicker enamel? Perhaps, but as paleoanthropologist Gen Suwa and his colleagues have pointed out, it may alternatively be that intermediate thickness is the primitive state, with humans and other African apes evolving away from it in opposite directions. Either way, *Australopithecus* and especially *Paranthropus* teeth have thick enamel. *Homo* teeth vary in enamel thickness, with early ones tending to be thinner than those of the robust australopiths in the same deposits. Humans have fairly thick enamel today but, as developmental biologist Tanya Smith has suggested, this may be a relative thing, a matter Harvard of less dentine rather than more enamel.

Truth be told, enamel thickness is a messy trait, and difficult to interpret. The thickness of the enamel cap depends on how and where it's measured, and it varies within and between species, and even within and between teeth of individuals. More importantly, there is likely no simple relationship between enamel thickness and diet in hominins. Strength and resistance to wear depend on how enamel is distributed across the crown, its microscopic structure, and its chemical composition. And thinness needs to be considered too. In fact, evolving thin enamel in strategic places can lead wear to sculpt a surface and form sharp edges for cutting tough items.

Foodprints. So far we've painted a picture of increasingly strong teeth through the first half of hominin evolution, well suited to heavy chewing, particularly of harder foods. Around 2.5 mya there was evidently a fork in the evolutionary road as savannas spread across eastern and southern Africa. The trend toward adaptation for heavy chewing continued with *Paranthropus*, but not *Homo.* Perhaps our *Homo* ancestors had higher-quality diets including meat, or they moved to processing foods outside the mouth with tools and, ultimately, fire, or both.

However, it is important to remember that tooth size, shape, and structure can tell us something about what an animal is capable of eating—but not what it actually eats on a daily basis. Real-world form–function relationships are much more complicated. African mangabey monkeys, for example, also have thick tooth enamel, flat teeth, and powerful jaws. Primatologists Scott McGraw and Joanna Lambert have each spent decades studying them, Scott in the Taï National Park, Ivory Coast, and Joanna in the Kibale National Park, Uganda. The Taï mangabeys specialize in very hard *Sacoglottis* nuts, about the size of a walnut, foraged from the forest floor. So far, so good. But the Kibale mangabeys prefer soft fruits, like other monkeys with less specialized teeth. That said, Kibale mangabeys still fall back on harder items, such as bark and seeds, when favoured foods are unavailable. Their specialized teeth give them more options and an advantage during hard times when choice foods are scarce. Another example comes from my own work. Orangutans have thicker enamel than other apes and monkeys that live alongside them in the Gunung Leuser National Park, Indonesia. They all eat large *Gnetum* fruits, which harden as they ripen. But the orangutans are able to eat the hardened fruits long after the other primates in the park are forced to abandon them. Thick enamel gives orangutans an edge.

So, how can we know whether adaptations reflect food preferences, fallback strategies, or something else? We can look to foodprints. As we've already discussed, carbon and other elements in teeth depend on the food an animal ate during tooth formation, and the microscopic use-wear scratches and pits on teeth were caused as items were pressed into or dragged along the enamel surface during chewing. Such evidence gives us important clues to the diets of animals alive in the past.

Recall that plants use light to transform carbon dioxide and water into carbohydrate and oxygen in different ways, so that they differ in their proportions of forms, or isotopes, of carbon (the isotopes of interest to us are ^{12}C and ^{13}C). Most tropical grasses and sedges (called C_4 plants) are made with more ^{13}C relative to ^{12}C than are trees, bushes, and shrubs (C_3 plants). And these isotope ratios are passed on to the animals that eat these plants. Tropical grass and sedge feeders have higher ratios of ^{13}C to ^{12}C in their tooth enamel than do animals that eat tree and bush parts. Early hominin carbon isotope ratios vary by species. *Ardipithecus* has a low ratio, suggesting a predominantly C_3 plant diet. *Australopithecus* is all over the map, with isotope ratios for different species including C_3 plant eaters, C_4 plant consumers, and mixed feeders. *Paranthropus* species also vary, from a C_4 plant diet to a more mixed one. Finally, early *Homo* individuals tend to have a mixed signal suggesting a broad diet including both plant types.

Recall also that microscopic tooth wear, or microwear, depends on the interaction between opposing teeth, and between teeth and abrasives in foods. Hard-object feeders tend to have heavy microwear pitting on their molars, whereas tough-food eaters have more scratches. Mixed feeders have both. Patterns of microwear in hominins also vary by species. *Ardipithecus* has wispy scratches, suggesting a softer- or tougher-food diet—things such as fruit pulp and leaves. *Australopithecus* also has mostly wispy scratches, but this varies by species. One *Paranthropus* species has wispy scratches, like the earlier hominins, but another is variable, with some individuals evincing heavy pitting. The latter is the pattern of the Kibale mangabeys, which, again, fall back on hard objects but prefer soft fruit when they can get it. Early *Homo* species, especially *Homo erectus* starting about 1.9 mya, also have variable microwear, but without the extreme pitting of some *Paranthropus* (see Figure 28).

When we assemble all the evidence, a few patterns emerge. First, it looks like *Ardipithecus* had a woodland diet of fruits and leaves. Gracile and robust australopith species had stronger teeth and experimented with other diets. Some were generalists, and took a broad range of foods in both savanna and forest, most often soft or tough, but at least one *Paranthropus* species also ate hard objects,

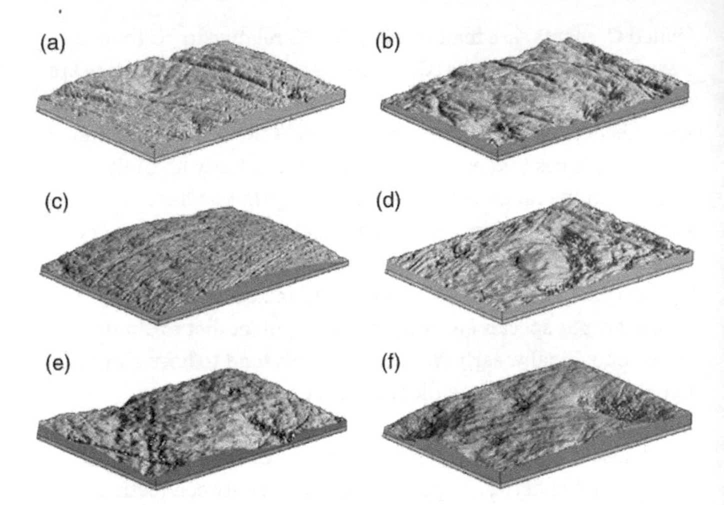

28. Dental microwear of *Australopithecus* (A, B), *Paranthropus* (C, D), and early *Homo* (E, F). Each image represents an area 0.1 × 0.14 mm

such as nuts or roots. Others were specialists, apparently eating mostly soft or tough grass or sedge parts. *Homo* species developed smaller, wimpier teeth over time, but had broad diets, judging from both carbon isotope and microwear evidence.

Evolutionary dentistry

A few hominin fossils show evidence of tooth decay and periodontal disease, just as some monkeys and apes have today, but these don't seem to have become rampant in our ancestors until very recently. And orthodontic problems, such as crooked or impacted teeth, were also rare in the distant past. Why are dental disease and orthodontic disorders so prevalent today? Dental paleopathologists address this question from an evolutionary perspective. They see it as a mismatch between our teeth and jaws on the one hand, and our diet on the other. In effect, our diet is changing too fast for our teeth and jaws to keep up. It's natural selection in action, at least for those unlucky enough to lack proper oral care or access to dental practitioners.

Dental caries. When plaque bacteria break down carbohydrates, they produce acid as a by-product. A drop in pH at the tooth's surface results in the loss of mineral and, ultimately, in dental caries, or progressive decay of enamel and dentine. While this affects about 90 per cent of young adults in the United States, little more than a handful of early hominin teeth have cavities. And few early modern humans had them. By some estimates, less than 2 per cent of Stone Age foragers had dental caries. Rates are also low in peoples that continue to hunt and gather wild foods for a living today. There are a few exceptions. The prehistoric lower Pecos hunter-gatherers of southern Texas and northern Mexico, for example, had terrible dental disease, probably due to a wild food diet rich in carbohydrates, which fed the plaque bacteria that cause dental decay (see Figure 29).

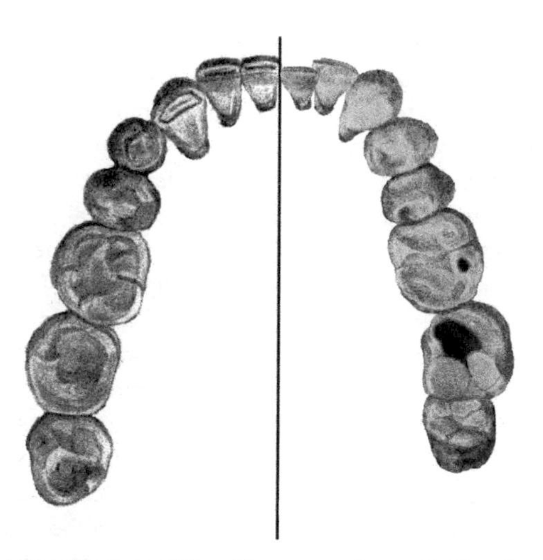

29. Stylized teeth of a traditional hunter-gatherer (left) and industrial-age person (right)

Following the Neolithic revolution—the invention and spread of agriculture—carbohydrate consumption surged as humans began to grow cereal grains. Along with this, caries rate increased something like fivefold. And it skyrocketed further in the 19th and 20th centuries with widespread availability of processed sugars and sugar-rich foods. Plaque bacteria break down sugar much more rapidly than they do other carbohydrates. This means more acid and more rapid tooth decay. There are certainly other factors to consider, such as genetic propensity, developmental defects, and pathological saliva. But diet change that came with the Neolithic and Industrial Revolutions clearly played a key role in increasing the rate of tooth decay. Indeed, when I asked John Sorrentino, a friend and dentist in New York, to recommend a toothbrush, he suggested that I worry more about my carbohydrate intake. That said, you should still brush your teeth—not only does it remove plaque bacteria that cause demineralization, but the fluoride can actually help remineralize enamel if the damage isn't too severe.

Periodontal disease. Oral bacteria also cause periodontal disease. Most of us have chronic gingivitis, an infection resulting in lesions, or wounds to our gums. This is the most common disease in the world today. And more than half of adults have at least some periodontitis, which involves damage to connective tissue and bone that support the teeth. Periodontal diseases are autoimmune disorders. The bacteria that form plaque release toxins that cause our immune systems to fight back. In response, we produce infection-fighting molecules called cytokines. While many causes and risk factors are associated with periodontal disease, excess cytokines are a major contributor to inflammation and tissue damage. Our immune response also produces white blood cells, which release an enzyme that breaks down connective tissues. Our own immune system is attacking our gums, the bone lining our tooth sockets, and the ligament that connects tooth and jaw. No wonder periodontal disease is the leading cause of tooth loss.

But periodontal disease is difficult to document in fossils and even recent skulls because similar damage to bone can occur during burial or after. And the earliest stages of disease may not even affect the jawbone. Still, we do occasionally see evidence of it, and tooth loss, in fossil hominins and early modern humans. Some modern hunter-gatherers with carbohydrate-rich diets also suffer periodontitis. It appears then that food plays a role in the rate of periodontal disease too. Early farmers in the New World, for example, tended to have a higher rate than did their hunter-gatherer predecessors, and industrialized people today seem to have a higher rate still. Nevertheless, the effects of the Neolithic and Industrial Revolutions on periodontal disease rate are not as clear as they are for caries, and much remains to be done to work through the causes and consequences.

Orthodontic disorders. Associations between modern diets and orthodontic disorders are clearer. Crowded, crooked, misaligned, and impacted teeth are huge problems today. These not only have important aesthetic implications, they can also decrease efficiency of food breakdown, lead to increased tooth decay, and compromise anchoring of the teeth in the jaw. Nine out of ten of us have at least slight malocclusion, and about half could benefit from orthodontic treatment. Like caries and perhaps periodontal disease, orthodontic disorders were much less common in fossil hominins and early peoples than they are today. And the change can come as quickly as one generation; we see it in the children of traditional foragers when they adopt a Westernized diet.

The problem is a mismatch between jaw length and tooth size. This has led to severe dental crowding at both ends of the tooth row. Many of us don't have enough room in our jaws for back teeth. Wisdom tooth impaction occurs ten times more often in modern society than in traditional hunter-gatherers. Our lower front teeth tend to be crooked and crowded together, and our uppers are pushed forward. Fossil hominins, early modern humans, and recent foragers more often had an edge-to-edge bite between opposing incisors rather than tips of the uppers resting in front of the lowers—thought by most clinicians to be normal occlusion.

Why the mismatch between jaws and teeth? In the 1920s, orthodontist Percy Raymond Begg discovered that prehistoric native Australians tend to have little malocclusion, but very worn teeth. He focused on approximal wear, which forms at contact points between adjacent teeth in a row as they rub together. Begg reasoned that teeth drift forward in the jaw to close the gap between them, and that jaw length is matched to worn tooth length. Maybe, then, our jaws are cramped because we don't wear our teeth enough. But as dental anthropologist Robert Corruccini has argued, maybe it's not our teeth that are too big, but our jaws that are too small. And indeed, human jaws have become shorter since the Early Stone Age. Our jaws are most likely underdeveloped because soft, highly processed foods don't provide the strain from heavy chewing needed to stimulate normal growth of the jaw during childhood. Remember Dan Lieberman's studies of jaw length and diet in laboratory animals? While I do not recommend that our children spend their days chewing on old boot leather, it's fun to imagine how much we could save on orthodontics bills if they did.

Chapter 8
Endless forms

> There is grandeur in this view of life, with its several powers, having been originally breathed into a few forms or into one; and that, whilst this planet has gone cycling on according to the fixed law of gravity, from so simple a beginning endless forms most beautiful and most wonderful have been, and are being, evolved.
>
> Charles Darwin, 1859

Teeth are important to me because they make the case for evolution. Endless forms most beautiful and most wonderful. The right teeth have given countless animals an edge in life's struggle for existence. The need for energy to survive and reproduce provides a powerful incentive. As prey animals and plants develop tough or hard tissues for protection, predators must evolve ways to sharpen or strengthen their teeth to overcome those defences. And a consumer must do a better job than its competitors or risk joining the innumerable ranks of evolutionary dead ends. Evolutionary biologist Lee Van Valen envisioned the process as an arms race between co-evolving species. As the Red Queen in Lewis Carroll's *Through the Looking-Glass* said, 'It takes all the running you can do, to keep in the same place.'

At the same time, meteors strike, continents shift, volcanos erupt, and climates vary with the Earth's tilt and orbit around the Sun.

These transform our world and change the foods available to hungry vertebrates in line at the biospheric buffet. And because carbohydrates, proteins, and lipids can all fuel the body, there's always something fresh with which to fill a plate. Living things must constantly adapt and change to survive when pitted against ever-evolving opposing organisms and an ever-changing world. And teeth must constantly adapt and change along with them. There is no stronger motive for the origin, evolution, and diversity of teeth today.

Of course, animals don't consciously evolve new teeth to face changing challenges of food acquisition and processing. And while natural selection is inevitable as conditions shift and rivals compete, the particular evolutionary path a species takes is not. It is easy to forget this when we trace the journey backward from today's moment in time. But the fossil record reminds us. There were so many interesting experiments in tooth form, so many starts and stops. Witness the razor-sharp oral plates of the Paleozoic armoured fish, *Dunkeosteus*, the elaborate dental battery of the Mesozoic dinosaur, *Hadrosaurus*, and the intricately carved teeth of the Cenozoic giant armadillo, *Glyptodon*. These and other examples demonstate how Nature works, and its endless possibilities.

New teeth can also create new opportunities. Recall that the earliest mammals were able to spread farther and into colder places because chewing allowed them to squeeze the energy needed to fuel an internal furnace. New habitats meant new potential resources, which fed back into selective pressures for even more new teeth. Also remember the appearance of the game-changing tribosphenic molar. The combination of shearing and grinding gave early mammals many options, and the versatility needed for a flexible diet in an unpredictable world. This also gave Nature a starting point from which all the subsequent myriad forms could evolve, from the simple peg-like cheek teeth of sloths and dolphins to the ornate molars of elephants and hippopotamuses.

This brings up the subject of evolvability. The ease with which teeth, especially mammalian teeth, can change is remarkable. Some tooth types must be very easy to make because they show up over and over again in unrelated species. Recall the crescent-shaped crested molars of koalas and cows, the sharp, V-shaped bladed cheek teeth of Tasmanian devils and lions, and the ever-growing, chisel-like incisors of rodents and vicuñas. And the same features also appear and disappear again and again. Hypocones come and go through evolutionary history like aeroplanes in and out of Heathrow airport. Other types are rare, but still show us how pliable teeth can be over evolutionary timescales. Think of the intricately folded bands of enamel on horse molars, the bizarrely hooked cheek teeth of the crab-eater seal, and pronged front teeth of colugos and grooved incisor of the solenodon, which forms a syringe for injecting venom. New research on evolutionary developmental biology is beginning to teach us how these things develop. A few drops of signalling protein is all it takes to add whole new cusps and other adornments to teeth grown in petri dishes.

Forward from the past

These are indeed exciting times for dental researchers. There are so many questions to tackle. How are teeth made? Why are they made as they are? How do animals use them? How can we use new knowledge about teeth and their evolution? The pace of discovery continues to accelerate as more and more scientists come at these questions from more and more directions.

How are teeth made? Evolutionary developmental biology is revolutionizing our understanding of how teeth are made and, more specifically, how genes signal embryonic cells to proliferate and differentiate into teeth. Are some types of teeth 'easier' to make than others? Does this explain why some shapes and structures show up again and again, whereas others do not? These questions are important not just for understanding why species respond to dietary needs as they do, but also for determining whether some similarities are better than others for inferring relatedness between extinct species. Researchers are hard at work in laboratories filling petri dishes with tooth germs to address these issues.

Why are teeth made as they are? New technologies are allowing researchers to document dental microstructure in unprecedented detail. A synchrotron particle accelerator, for example, can produce X-rays bright enough to make a 3D model of the inside of a tooth with a resolution less than a thousandth of a millimetre. The layout of tissue structure on this level can teach us how teeth resist and dissipate the stresses that come with chewing. When we combine this knowledge with understandings of how teeth break food, we can better address the question 'Why are teeth made this way?'

Researchers are also using cutting-edge tools to document tooth structure at larger scales. X-ray microtomography lets us map the distribution of enamel across a tooth crown in 3D. Is enamel especially thick over the cusps to strengthen the tooth for crushing hard objects? Is it thin to wear through quickly to dentine horns and create sharp edges for shearing tough foods? We are just now starting to figure out how Nature uses wear for sculpting occlusal surfaces to make and keep them the best shapes possible for their given tasks.

How do animals use teeth? Studies of how foods with different properties fracture allow us to make idealized models of the best tools for breaking them. When we compare these models to real teeth, though, tooth shape is not always what we expect given diet, and diet not always what we expect given tooth shape. Researchers are working hard to improve their models and explain the apparent discrepancies. For example, it is difficult to believe that sloths and koalas, or pandas and bamboo lemurs, have such similar diets given differences between their teeth.

As we move from models to real life, it becomes clear that Nature does what it can with the raw materials available. Distantly related species can converge on a given diet from different morphological starting points. Their heritage gives them what we call *phylogenetic baggage* to deal with; they start burdened with the tooth shape they inherit. Imagine a landscape of fitness in which better adapted forms sit at higher elevations. If you keep moving uphill you'll eventually reach a peak, but it may not be the highest in the range. It all depends on where you start. If the next peak over is higher, you can't reach it without descending through the valley between. Nature tends to push upward, not down. So a sloth, for example, is stuck with the dentine pegs it inherited from its ancestor. Without understanding this, you'd probably never guess from its teeth that the sloth is a leaf eater. This is the old function versus phylogeny problem that keeps paleontologists up at night. Those of us that work with fossil teeth must develop ways of separating diet from heritage or we have little hope of deciphering the past.

Even when form-to-function relationships are clear, the selective pressures driving them may not be. Mountain gorillas have sharp teeth, heavy jaws, and massive chewing muscles. At Karisoke in the Virunga Mountains of Rwanda, gorillas eat mostly tough, fibrous plant parts. Here anatomy matches behaviour. But the Karisoke apes don't have much of a choice; there is little other food at higher altitudes, and humans have settled the valleys. Gorillas at lower elevations in the nearby Bwindi Forest of Uganda more often eat soft, sugary fruit. Still, their ability to eat leaves and other plant parts gives them an advantage when and where favoured foods are unavailable. The selective pressures and teeth are the same for higher- and lower-elevation gorillas, but their daily diets are not. In this case, choice is all about what's offered on the biospheric buffet. Studies of animals in the wild give us a better understanding of how natural selection works, and continued research on feeding ecology and foraging strategies in light of geographic, seasonal, and longer-term fluctuations in food availabilities will undoubtedly lead to new insights.

As we have seen, foodprints can help us to extract information from fossil teeth. Stable isotope researchers are today working out the chemical signatures of different foods, and how these end up in the teeth of animals that eat them. Also, microwear researchers are documenting the patterns of tooth wear that different sorts of foods cause. Such evidence, when combined with tooth shape and morphological starting point, can help us reconstruct diets of past animals.

How can we use this knowledge? Basic research has intrinsic worth. It expands human knowledge and satisfies our curiosity. But it also lays the foundation for applied research to solve real-world problems. And there is practical value in knowing how teeth develop and evolve. Engineers are beginning to realize that Nature has been tinkering with and improving tooth structure and form for nearly half a billion years. A better understanding of teeth can lead to bio-inspired designs for self-sharpening tools, and all kinds of structures that require strength and durability.

Research on dental development and evolution also has important clinical implications. We are learning more and more about how to make a tooth, and some believe that regenerative therapy is right around the corner. Can we bioengineer a new tooth rather than use prosthetic crowns, implants, or dentures to deal with damage, decay, or loss? Time will tell. In the meantime, an evolutionary perspective can certainly inform clinical research and practice. To be sure, better oral hygiene, fluoridation, and dental care help in prevention and treatment, but that's still a far cry from the oral environments in which our teeth evolved. Dental crowding due to the mismatch between our jaw and tooth sizes provides a great example. Studies of past peoples reveal that our jaws are too small for our teeth, rather than the other way around. Wouldn't it make sense, then, for orthodontists to focus more on lengthening the jawbone and less on reducing tooth mass through extraction and reshaping?

And there are many other applications. As but one example, fossil teeth can help us understand the long-term effects of climate change on life. Paleoclimatologists and paleontologists are working together to match fluctuations in temperature and precipitation over deep time with extinctions and evolutionary events recorded in the fossil record. Since climate affects local environment, which in turn determines food availability and diet, changes in tooth form can help us understand how species reacted to climate change in the past. This can help us predict how they will respond in the future. Researchers are digging deep to fill gaps in the fossil record, both to better document evolution and to tie it to environmental dynamics.

We still have much to learn about teeth, how they evolve, develop, are put together, and used. An evolutionary approach can help us understand our legacy, and guide us forward.

Further reading

There are countless thousands of scientific papers and many excellent books on teeth. Here I present just a few to get the reader started.

Encyclopedic treatment of teeth

P. S. Ungar, *Mammal Teeth: Origin, Evolution, and Diversity* (Baltimore: Johns Hopkins University Press, 2010).

E. Thenius, *Zähne und Gebiß der Säugetiere* (Berlin: Walter de Gruyter Press, 1989).

B. Peyer, *Comparative Odontology* (Chicago: University of Chicago Press, 1968).

C. G. Giebel, *Odontographie: Vergleischende Darstellung des Zahnsystemes der Lebenden und Fossilen Wirbelthiere* (Lepizig: Verlag von Ambrosius Abel, 1855).

R. Owen, *Odontography* (London: Hippolyte Baillierc, 1840).

Dental development and microstructure

M. Bath-Balogh and M. J. Fehrenback, *Illustrated Dental Embryology, Histology, and Anatomy*, 3rd edition (St Louis: Saunders, 2010).

A. Nanci, *Ten Cate's Oral Histology: Development, Structure, and Function*, 8th edition (St Louis: Elsevier Mosby, 2013).

M. F. Teaford, M. M. Smith, and M. W. J. Ferguson, *Development, Function, and Evolution of Teeth* (New York: Cambridge University Press, 2000).

Functional morphology and biomechanics of teeth

T. Koppe, G. Meyer, and K. W. Alt, *Comparative Dental Morphology* (Basel: Karger, 2009).

P. W. Lucas, *Dental Functional Morphology: How Teeth Work* (New York: Cambridge University Press, 2004).

P. Smith and E. Tchernov, *Structure, Function, and Evolution of Teeth* (London and Tel Aviv: Freund, 1992).

Paleontological treatments of teeth

N. Shubin, *Your Inner Fish: A Journey into the 3.5-Billion-Year History of the Human Body* (New York: Vintage Books, 2009).

K. D. Rose, *The Beginning of the Age of Mammals* (Baltimore: Johns Hopkins University Press, 2006).

T. S. Kemp, *The Origin and Evolution of Mammals* (Oxford: Oxford University Press, 2005).

P. Janvier, *Early Vertebrates* (New York: Oxford University Press, 2003).

Paleoanthropological and archeological treatments of teeth

J. D. Irish and G. C. Nelson (editors), *Technique and Application in Dental Anthropology* (New York: Cambridge University Press, 2008).

S. E. Bailey and J.-J. Hublin (editors), *Dental Perspectives on Human Evolution: State of the Art Research in Dental Paleoanthropology* (Dordrecht: Springer, 2007).

P. S. Ungar (editor), *Evolution of the Human Diet: The Known, the Unknown, and the Unknowable* (New York: Oxford University Press, 2007).

S. Hillson, *Teeth*, 2nd edition (Cambridge: Cambridge University Press, 2005).

S. Hillson, *Dental Anthropology* (Cambridge: Cambridge University Press, 1996).

K. W. Alt, F. W. Rösing, and M. Teshler-Nicola (editors), *Dental Anthropology: Fundamentals, Limits, and Prospects* (Vienna: Springer, 1998).

Evolutionary dentistry

R. S. Corruccini, *How Anthropology Informs the Orthodontic Diagnosis of Malocclusion's Causes*. (Lewiston: Edwin Mellen Press, 1999).

Index

A

B

C

D

I

J

K

L

M

S

T

U

V

W

X

Z